Quantum Artificial Intelligence

Vijayarangan Natarajan
Editor

Quantum Artificial Intelligence

A Machine-Generated Literature Overview

Springer

Editor
Vijayarangan Natarajan
IIT Madras Research Park
Tata Consultancy Services Limited (TCS)
Chennai, Tamil Nadu, India

ISBN 978-981-96-5050-7 ISBN 978-981-96-5051-4 (eBook)
https://doi.org/10.1007/978-981-96-5051-4

This Springer imprint is published by the registered company Springer Nature Singapore Pte Ltd.
The registered company address is: 152 Beach Road, #21-01/04 Gateway East, Singapore 189721, Singapore

Preface

The field of quantum artificial intelligence (QAI) deals with solving academia and industrial challenges and improving the experience. This field involves multiple subjects like quantum information science, quantum computing, quantum programming, post-quantum cryptography, simulation, and quantum machine learning. It involves a lot of algorithms, computing methods, quantum spin, quantum particles, devices, and experiments toward analyzing the challenges.

This book, *Quantum Artificial Intelligence*, describes various research papers of QAI arising out of the challenges. It is a collection of research articles pertaining to QAI focusing on fundamental, programming, computational, and functional applications and challenges. There are six chapters presented in this book: Quantum information processing, Quantum programming, Post-quantum cryptography, Simulation of quantum and artificial systems, Quantum machine learning, and Applications. Each chapter is written based on theoretical and practical problems and solutions serving the unsolved problems related to academia and industry.

This book contains a variety of interdisciplinary academic fields and articles; the readers should have domain knowledge, including fundamentals of artificial intelligence, cryptography, programming, machine learning, mathematics, physics, mechanics and Computational, Artificial intelligence and Machine learning Systems (CAMS). This book is intended for graduate students, research scholars, academicians, and industry professionals.

The author of this book is grateful to God, family members, colleagues, friends, and publisher who encouraged a lot to investigate in a new era of QAI with a fathom of learning. This book is dedicated to them.

A special thanks to Shrutika Dasari of Springer Nature for editing the entire manuscript.

As written by the author: VIJAYARANGAN NATARAJAN

Dr. Vijayarangan Natarajan obtained his Ph.D. (Mathematics) in the year of 2001 at the Ramanujan Institute for Advanced Studies in Mathematics, University of Madras, Chennai. His doctoral research work was on Krein H*, J*-algebras, and triple systems. His research work includes mobile computing, Hilbert algebras, Jordan algebras, Lie algebras, number theory, elliptic curve cryptography, communication protocols, quantitative analysis, applications of real and complex analysis in image processing, artificial intelligence, machine learning predictions, stochastic computing, etc. He has also received the *Best Research Paper Award* given by the Ramanujan Mathematical Society in June 2000. He has a rich flavor of academic and industry experience around 30 years. He has presented his research work at several universities globally and published more than 100 articles, including papers, patents, and books. He is keen to focus on research areas like artificial intelligence, quantum computing, and stochastic process. He is fond of physical fitness too.

Chennai, Tamil Nadu, India — Vijayarangan Natarajan

About This Book

Auto-summaries can be generated by either an abstractive or extractive auto-summarization:

- An extraction-based summarizer identifies the most important sentences of a text and uses the original sentences to create the summary.
- An abstraction-based summarizer creates new text based on deep learning. New phrases are created to summarize the content.

The auto-summaries you will find in this book have been generated via an extractive summarization approach.

Each chapter was carefully edited by Dr. Vijayarangan Natarajan. The editors selected the papers which were then auto summarized. The editors have not edited the auto-summaries due to the extraction-based approach and have not changed the original sentences. You will find the editors' reviews and guidance on the auto-summaries in their chapter introductions.

In machine-generated books, editors are defined as those who curate the content for the book by selecting the papers to be auto-summarized and by organizing the output into a meaningful order. Next to the thoughtful curation of the papers, editors should guide the readers through the auto-summaries and make transparent why they selected the papers.

The ultimate goal is to provide a current literature review of Springer Nature publications on a given topic in order to support readers in overcoming information overload and to help them dive into a topic faster; to identify interdisciplinary overlaps; and to present papers which might not have been on the readers' radar.

Please note, that the selected papers are not used to train a LLM while the auto-summaries are created.

Introduction of the Book

The book *Quantum Artificial Intelligence* discusses the research on quantum information science, computational intelligence, quantum algorithms, machine learning, deep learning, artificial intelligence, and applications. This book covers quantum computing and quantum devices. To learn quantum artificial intelligence, it requires a variety of subjects like combinatorial optimization, soft computing, neural networks, learning algorithms, simulations, information science, and quantum mechanics.

Quantum computing is a growing subject which has an interrelationship with artificial intelligence, the computer science researchers explored. In this book, we provide the viability of practical applications in many domains. It is helpful to academia and industry to learn quickly about quantum artificial intelligence.

This book covers six chapters such as Quantum information processing, Quantum programming, Post-quantum cryptography, Simulation of quantum and artificial systems, Quantum machine learning, and Applications. Each chapter covers the in-depth knowledge of its domain. Chapter 3, Post-quantum cryptography, has some new crypto algorithms involving high-end computations. Finally, we have covered sufficient applications of quantum computing to gain a paramount of wisdom in this field.

Contents

Chapter 1
Quantum Information Processing

Introduction by the Editor

This chapter deals with an estimation theory of quantum particles and systems. The estimation of all parameters has been encoded in quantum state and processed through a quantum information matrix. The researchers have used the Born-Markov approximation for quantum estimation. Quantum mechanics has proven to be a valuable resource for the investigation of the behavior of complex networks. With the arrival of quantum computers, new algorithms have come out to do quantum chemistry, and importantly, atomic-like orbitals. Quantum compression algorithms help to develop the landscape of emerging IT technologies, IoT and Internet. Quantum Image processing (QIMP) has become a popular area of quantum research due to the ubiquity and primacy of digital image and video processing in modern life. In modeling noise in the quantum circuits, we assume a model by which each one- and two-qubit gates are followed by a depolarizing noise channel. Quantum computing (QC) potential use in High Energy Physics has led CERN, one of the top world users of large-scale distributed computing, to start programs such as the Quantum Technology Initiative (QTI) to further assess and explore the applications of QC. This chapter gives a survey on quantum cryptography in which quantum key distribution protocols are proven secure for industrial applications.

In this chapter, we outline a study involving quantum estimation and the dynamics of entangled qubits interacting with a bosonic environment. Then we cover up entangled qubits, bosonic bath interaction, Markovian master equation, quantum Fisher information and quantum correlation indicators. Further, we discuss the application of quantum-inspired measures for graph similarity to address challenges in graph comparison, which is relevant in various domains such as graph classification, outlier detection, and identifying interaction patterns in complex systems. This topic leads to highlight graph similarity and distinguishability, challenges in graph comparison, quantum-inspired measures for network similarity and experimental

V. Natarajan (ed.), *Quantum Artificial Intelligence*,
https://doi.org/10.1007/978-981-96-5051-4_1

evaluation. We bring out a procedure for defining virtual spaces and computing one-electron and two-electron integrals for plane-wave second quantized Hamiltonians in periodic systems. We cover up utilization of quantum computers through Microsoft Azure quantum service, correlation optimized virtual orbitals (COVOs), incorporation of first Brillouin zone integration and aperiodic systems.

In this chapter, we provide an overview of data compression algorithms within the context of edge computing, where processing is performed locally to minimize data transfer to centralized data centers. Edge computing addresses challenges associated with transferring large volumes of data generated remotely to centralized data centers efficiently.

We highlight the transformative potential of quantum technology and emphasize the importance of anticipating its future implications, including ethical, legal, social, and policy considerations. Then we describe the role of quantum computing (QC) in High Energy Physics, particularly at CERN, and report on the design and delivery of a series of lectures on quantum computing as part of the Quantum Technology Initiative (QTI).

Next, we focus on quantum renewable energy which shows the potential for quantum computing to address challenges in renewable energy, particularly in the context of the electrical grid.

In this chapter, we delve into the intersection of quantum information theory and relativistic quantum mechanics. The research demonstrates that in relativistic settings, the evolution of a massive spin-1/2 particle can contravene several standard assumptions of quantum information theory. Further, quantum physics explores the application of novel concepts surrounding uncommon quantum states to the physics of condensed matter, particularly focusing on solids within the framework of contemporary field theory. Also, we address the challenge of preserving entangled states in the presence of environmental effects, such as decoherence and dissipation, which can degrade the quantum information stored in a system. We introduce the concept of quantum programming, highlighting its interdisciplinary nature, drawing on principles from quantum mechanics, mathematics, and computer science. It sets the stage for understanding the significance of quantum programming languages in the context of quantum computing.

Quantum programming covers aspects such as support for quantum primitives and operations, quantum circuit construction, quantum algorithm implementation, and integration with classical programming languages.

The research survey exploits new ideas and methods in quantum states applied to condensed matter physics, especially solids, within contemporary field theory. It compares modern approaches with classical many-electron theory, focusing on ground states, excitations, quantum fluctuations, and phase transitions, with a particular emphasis on topological aspects and frustration effects. The survey covers variational methods, mean-field approximations, gauge field theory, and exotic states, alongside an overview of the theory of entangled topological states, spin liquids, and string networks.

In this chapter, we present how physical implementations of cryptographic algorithms can leak information, making them susceptible to side-channel attacks. The concept of secure computation amid such leakage is termed leakage resilience. An innovative research brings out leakage resilience with fault-tolerant quantum computation, demonstrating that a noise model ensuring fault tolerance also implies leakage resilience. For fault-tolerant implementation at the quantum gate, the researchers apply Toffoli gate in order to perform quantum classical translation. Further, the researchers have shown that leakage-resilient gadgets construction in the quantum circuits will be executed through fault-tolerant implementation.

The research work "Stationary States of a Dissipative Two-Qubit Quantum Channel and Their Applications for Quantum Machine Learning" describes the behavior of two-qubit systems under dissipation. It shows how stationary states can be harnessed for quantum machine learning or deep learning applications. The research study illustrates that by managing dissipation, it is possible to stabilize entangled states, which are crucial for quantum computing and machine learning algorithms.

The authors explore various methods to control and utilize dissipative processes to maintain coherence and entanglement in two-qubit systems. These methods are applied to create robust quantum states that can be used in quantum machine learning tasks, providing a foundation for developing more efficient quantum algorithms and improving the reliability of quantum computations in the presence of environmental interactions. The recent research work describes how to manage these adverse effects to provide stationary entangled states for quantum machine learning (QML). Further, the research brings out that the channel can determine if an initial quantum state will result in a stationary entangled state. This algorithm is applied to propose a theoretical framework for quantum neural networks (QNNs) using variational quantum circuits. Then these circuits encode data in the continuous variables (CVs) of the two-qubit states. Subsequently, linear, and nonlinear transformations in the QNN are implemented using the stationary states of the two-qubit channel and the measurement process, respectively. The model's effectiveness is tested on supervised binary classification tasks, demonstrating that integrating non-unitary transformations and parallel-processed neural computing in such a channel meets the requirements for a meaningful QNN. This CV-QNN model, with enough layers, could execute any algorithm feasible on a universal CV quantum computer. Nowadays, a lot of research programs fulfill the challenges in qubit channels and the measurement process through QML.

At the end of Chap. 1, we bring out a significant connection between two important areas of research: leakage resilience in cryptographic algorithms and fault-tolerant quantum computation. By leveraging techniques from quantum error correction and fault tolerance, the research provides insights into how cryptographic schemes can be made resilient to leakage.

Machine Generated Summaries

Disclaimer: The summaries in this chapter were generated from Springer Nature publications using extractive AI auto-summarization: An extraction-based summarizer aims to identify the most important sentences of a text using an algorithm and uses those original sentences to create the auto-summary (unlike generative AI). As the constituted sentences are machine selected, they may not fully reflect the body of the work, so we strongly advise that the original content is read and cited. The auto generated summaries were curated by the editor to meet Springer Nature publication standards. To cite this content, please refer to the original papers.

Machine generated keywords: spin, theory, initial state, edge, topological, survey, material, interaction, article, mathematical, electronic, basis, quantum compute, ground state, series.

Multiparameter Estimation for a Two-Qubit System Coupled to Independent Reservoirs Using Quantum Fisher Information [1]

This is a machine-generated summary of:

Bukbech, S.; El Anouz, K.; El Allali, Z.; Metwally, N.; El Allati, A.: Multiparameter estimation for a two-qubit system coupled to independent reservoirs using quantum Fisher information [1].

Published in: Quantum Studies: Mathematics and Foundations (2023).

Link to original: https://doi.org/10.1007/s40509-023-00303-6

If you want to cite the papers, please refer to the original.

For technical reasons we could not place the page where the original quote is coming from.

Abstract-Summary

"The quantum Fisher information is used as an estimator of the phase damping of the initial state settings and the interaction damping parameter."

"It is exhibited that the initial state settings play a role as a controller of the precision of the estimation degree of these parameters."

"We classified the state that could be used to estimate the phase damping with high accuracy."

"The constant behavior of the variances of the estimated parameters shows that the optimal estimation could be achieved for small values of the damping parameter."

Introduction

"Various outstanding results in statistical theory and quantum information theory have been explained according to the concept of quantum measurement, which gives rise to many substantial theories, including quantum metrology, quantum estimation theory and so on. The major goal of quantum estimation theory is to perform robust precision measurements for parameters included in quantum systems."

"In this spirit, improvements in precision limits can be fulfilled with the simultaneous quantum estimation of all parameters encoded in the quantum state [2–4]."

"Exhaustively, the process of estimating various parameters at the same time provides more robust precision than estimating them individually via Quantum Fisher Information Matrix (QFIM)."

"As we have already pointed out, the QFIM is the key quantity representing the precision limit for multiparameter quantum estimation because of its applications in different domains [5–12]."

From Classical to Quantum Estimation Theory

"One of the most significant approaches to multiple parameter estimation is to perform individual measurements for each parameter."

"Estimating all parameters simultaneously provides robust precision bounds [4]."

"We examine some preliminaries of the multi-parameter quantum estimation theory, where we shall investigate the QFIM and construct the corresponding QCRB."

"The QFIM is a fundamental tool on which multiparameter quantum estimation is built."

"The estimation precision bound is determined by the quantum multiparameter QCRB."

"Saturating the QCRB is a critical issue in quantum estimation."

"For quantum single-parameter estimation, the QCRB in principle can be reached all the times [4]."

"The problem arises in multi-parametric quantum estimation, because quantum measurements are not always commutative."

"The non-existence of an optimal common measurement for the estimation of all parameters."

Proposed Model

"In the context of the Born–Markov approximation, the bath correlation is important in determining how the system evolves over time."

"In this inspiration, we should mention that in the above calculations, we have used the Born–Markov approximation as a good assumption that simplifies the derivation of the master equation by assuming that the environment relaxes to its equilibrium state after being perturbed by the system."

"We intend to verify whether we have an optimal estimation using the proposed model."

[Section 1]

"This indicates that the estimation of the phase parameter depends on the initial state settings, whether it is partially/maximally entangled."

"This means that the initial state is prepared in a maximally entangled state of the Bell type (MES)."

"The initial state settings have a major role in governing the decay rate, where those prepared in MES are the most sensitive to the damping parameter."

"The increasing rate depends on the initial state settings, whether it is a maximally or partially entangled state."

"This confirms that the initial state settings could be used as a controller on the precision of estimation."

"The ability to estimate the damping parameter, however, is dependent on whether the initial qubit system is prepared in a MES or a PES."

Conclusion

"The possibility of estimating the phase of the initial state and the damping parameters of the environment is discussed, where we estimate these parameters individually and simultaneously by using the quantum Fisher information (QFI) as an estimator."

"At a later time, where the initial state turns into a partially entangled state (PES), the estimation process of the phase parameter increases."

"The decay rate of QFI as an estimator of the phase parameter that is depicted for maximally entangled states is larger than the one displayed for partially entangled states."

"The sensitivity of the estimation degree to large values of the damping parameter is discussed for different initial state settings."

"The behavior of the minimum variance of the damping parameter shows that, one can estimate it with high precision if the initial system is prepared in a partially entangled state."

Quantum-Inspired Measures of Network Distinguishability [13]

This is a machine-generated summary of:

Polychronopoulou, Athanasia; Alshehri, Jumanah; Obradovic, Zoran: Quantum-inspired measures of network distinguishability [13].

Published in: Social Network Analysis and Mining (2023).

Link to original: https://doi.org/10.1007/s13278-023-01069-w

If you want to cite the papers, please refer to the original.

For technical reasons we could not place the page where the original quote is coming from.

Abstract-Summary

"The recent emergence of information physics as a theoretical foundation for complex networks has inspired the utilization of measures, initially developed for use with quantum mechanical systems, for the solution of graph theory research problems."

"A network similarity measure is required for any data mining application on graphs, such as graph clustering, classification, or outlier detection."

"A natural starting point for the identification of such a network similarity measure is information physics, offering a series of measures typically used to quantify the distance of quantum states."

Introduction

"A variety of network theoretic tools are used for the analysis of these complex systems (Bunke et al. [14]), and the question of graph distinguishability or graph similarity is often a central aspect of study."

"Quantum mechanics has proven to be a valuable resource for the investigation of the behavior of complex networks (Biamonte et al. [15])."

"Quantum gases have been used to describe network evolution, and the emergence of different structures in complex networks has been represented in terms of a quantum-classical transition for quantum gases (Javarone et al. [16])."

"These quantum-inspired methods offer a mathematical method of graph comparison, they satisfy the mathematical properties of a metric, offer intelligible results with high interpretability, and were shown in Polychronopoulou et al. [17] to satisfy intuitive graph similarity properties for monoplex networks."

Methods

"This formulation allows multiplex networks to be compared using the distance measures introduced for single layer graphs, with the supra-adjacency matrix substituting the monoplex adjacency."

"Alternative representations have been proposed for multiplex networks, typically constructing a monoplex network by aggregating the various layers."

"The authors of Sánchez-García et al. [18] introduced the notion of graph quotients and an aggregate network that summarizes the connections of nodes among the different layers, normalised by the multiplexity degrees of the nodes."

"If the monoplex aggregate is seen as the network describing the absolute connection between nodes, and the layers are seen as evidence of this connection, then the extension for use with multiplex networks is natural."

"The aggregate monoplex describes the reliability of connections between nodes, as affected by multiple sources of information, the layers of the multiplex network."

Evaluation on Artificial Data

"In the simple case of repeated node removals, the monotonically increasing behavior of the distance measures is evident and confirms that they all act as measures of graph distinguishability."

"With the exception of Bures distance, the measures have identified that the removal of one node is more crucial for smaller networks, as is evident by the slope of the curves."

"All distances are sensitive to distinguishing targeted from random operations, both for the case of continuous edge removals targeting popular nodes, and continuous edge removals targeting edges of higher weights."

"Each network is then continuously modified, applying at each step one elementary graph edit operation, and the distance between the modified and original network is calculated."

"To the case of monoplex networks, quantum distances satisfy intuitive similarity aspects: they continuously increase as the networks are further modified and this applies to most multiplex representations and all types of data."

Applications on Real Data

"As a first application, we evaluate the effectiveness of the distance measures on classification problems with real world single layer networks."

"Following this motivation, we evaluate the effectiveness of the quantum-inspired graph distance measures in the problem of computing pairwise similarities between layers, and then performing hierarchical clustering of structurally similar layers in multiplex networks."

"The plot indicates that most of the distances are able to achieve an accurate classification when two or more layers are used, with an F1 score that is increasing with the addition of more layers."

"What is evident is that traditional methods, and the baselines, such as DELTACON, as well as quantum distances are increasing their accuracy with the addition of more layers."

"For many of the multiplex representations the accuracy is starting to drop after the addition of a given number of layers, defining a point at which the complexity of the problem increases, more evidently for the edge-weight and trace distances."

Discussion

"These measures are well-established mathematical methods that incorporate the intrinsic structure of the entire network and have high interpretability."

"Quantum and information physics undoubtedly compose a rich resource of mathematically established and interpretable measures, used for the evaluation or comparison of natural systems."

"Many of these measures have been translated into meaningful complex network tools, in this work and previous ones (Biamonte et al. [15])."

"In the case of multiplex networks, quantum states separability can be seen as layer separability as well possible measure of the layers' informative power."

Periodic Plane-Wave Electronic Structure Calculations on Quantum Computers [19]

This is a machine-generated summary of:

Song, Duo; Bauman, Nicholas P.; Prawiroatmodjo, Guen; Peng, Bo; Granade, Cassandra; Rosso, Kevin M.; Low, Guang Hao; Roetteler, Martin; Kowalski, Karol; Bylaska, Eric J.: Periodic plane-wave electronic structure calculations on quantum computers [19].

Published in: Materials Theory (2023).

Link to original: https://doi.org/10.1186/s41313-022-00049-5

If you want to cite the papers, please refer to the original.

For technical reasons we could not place the page where the original quote is coming from.

Abstract-Summary

"A procedure for defining virtual spaces, and the periodic one-electron and two-electron integrals, for plane-wave second quantized Hamiltonians has been developed, and it was validated using full configuration interaction (FCI) calculations, as well as executions of variational quantum eigensolver (VQE) circuits on Quantinuum's ion trap quantum computers accessed through Microsoft's Azure Quantum service."

"This work is an extension to periodic systems of a new class of algorithms in which the virtual spaces were generated by optimizing orbitals from small pairwise CI Hamiltonians, which we term as correlation optimized virtual orbitals with the abbreviation COVOs."

"With these procedures, we have been able to derive virtual spaces, containing only a few orbitals, that were able to capture a significant amount of correlation."

"Calculations performed on the Quantinuum H1–1 quantum computer produced surprisingly good energies, in which the error mitigation played a small role in the quantum hardware calculations and the (noisy) quantum simulator results."

Introduction

"With the arrival of quantum computers, researchers are actively developing new algorithms to carry out quantum chemistry calculations on these platforms, in particular for calculations containing strong electron-electron correlations (aka high-level quantum chemistry methods)."

"Most high-level quantum chemistry methods in use today (e.g., full configuration interaction (FCI) (Szabo and Ostlund [20]; Ross [21]; Gan et al. [22]; McArdle et al. [23]; Tubman et al. [24]; Sugisaki et al. [25]; Kawashima et al. [26]), coupled cluster (CC) (Coester [27]; Coester and Kummel [28]; Čížek [29]; Paldus et al. [30]; Mukherjee et al. [31]; Purvis and Bartlett [32]; Bishop and Kümmel [33]; Paldus and Li [34]; Crawford and Schaefer [35]; Bartlett and Musiał [36]; Peng et al. [37]), and Green's function (GF) (Green [38, 39]; Feynman [40, 41]; Martin and Schwinger [42]; Baym and Kadanoff [43]; Peng et al. [37]) approaches) are based on second-quantized Hamiltonians, which are written in terms of the creation and annihilation operators of the Fermion orbitals along with the one-electron and two-electron integrals for the system."

"One of the first, and still popular, class of basis sets used in quantum chemistry methods are atomic-like orbitals or the linear combination of atomic orbitals (LCAO) basis set."

"We have recently developed new methods for generating optimized orbital basis sets, called COVOs (Bylaska et al. [44])."

Pseudopotential Plane-Wave Second-Quantized Hamiltonian

"A standard way to remove this issue in plane-wave calculations is to replace these singular potentials by pseudopotentials."

"For methods that utilize linear combinations of atomic orbitals (LCAO) as the basis, the size of the basis set and subsequently generated Hartree-Fock orbitals is fairly small."

"Many of the periodic forms presented in the following sections are written in terms of Fourier space using periodic plane-wave basis sets, rather than real space."

"Descriptions of the plane-wave methods used in this work can be found in the following references (Bylaska et al. [45, 46]; Bylaska [47]; Bylaska and Rosso [48]; Bylaska et al. [49]; Pickett [50]; Ihm et al. [51]; Car and Parrinello [52]; Payne et al. [53]; Remler and Madden [54]; Kresse and Furthmüller [55]; Marx and Hutter [56]; Martin [57]; Valiev et al. [58]; Chen et al. [59])."

Algorithm for Defining a Virtual Space with a Small CI Hamiltonian

"These new types of orbitals are able to capture significantly more correlation energy than with virtual orbitals coming from Hartee-Fock (Bylaska et al. [44])."

"The above formulas can be generalized to work beyond two electron systems by using corresponding orbitals techniques (King et al. [60]; Bylaska and Rosso [48])."

"We also note the COVOs approach is similar in spirit to the optimized virtual orbital space (OVOS) approach developed over 30 years ago by Adamowicz and Bartlett Adamowicz and Bartlett [61]; Adamowicz et al. [62]."

"The corresponding orbital transformation (King et al. [60]) can be used to generalize for different orthonormal sets."

"In cases, where the two-electron matrix elements of the spin-orbitals have a double noncoincidence (King et al. [60]) the matrix elements are again block diagonal, otherwise the matrix elements can be represented as a sum of periodic Coulomb and exact exchange energies, where the Filon integration strategy (Bylaska et al. [49]) can be used to fold in the first Brillouin zone integration present in the exact exchange energies."

Results for the Ground State of the LiH Molecule Using Periodic Boundary Conditions

"The NWChem program package (Kendall et al. [63]; Valiev et al. [64]; Bylaska et al. [45]; Bylaska [47]; Apra et al. [65]) was used for all calculations in this study, except for the FCI calculations, which used the TINYMRCC suite by Jiří Pittner."

"The average difference error for the 1, 4, 8, and 12 COVOs calculations from the 18 COVOs calculation is 12.9 kcal/mol, 2.7 kcal/mol, 1.0 kcal/mol, and 0.4 kcal/mol respectively."

"The energy for large R should be the same as the combined energy of the isolated H and Li atoms."

"At large R a significant difference between aperiodic and periodic calculations can be observed."

Quantum Computer Calculations for the Ground State of the LiH Molecule Using Periodic Boundary Conditions

"To computing ground-state energies at different bond lengths on a quantum computer, we also wanted to measure the effects of noise on the corresponding circuit evaluations."

"The VQE method is a hybrid quantum-classical approach in which energies are evaluated on quantum hardware or simulators, and classical computers perform the algorithm to optimize the variational parameters."

"Proposals for robust quantum error correction require qubit numbers and performance that are not yet available via Cloud-based NISQ devices today (Shor [66]; Cory et al. [67]; Reed et al. [68]), so before executing the circuits on the H1–1 quantum computer, we wanted to ensure that noise played a manageable role in computing the ground-state energies."

"After convincing evidence that the error from the noise for this circuit can be well tempered, we performed the last stage of the calculations, where the same energy evaluation and error mitigation technique was performed for 500 runs on the Quantinuum H1–1 quantum computer."

Conclusion

"For an (N + 1)-state Hamiltonian, the method is based on optimizing the virtual orbitals to minimize a small select CI Hamiltonian (i.e., COVOs) that contains configurations containing all N filled RHF orbitals and the one virtual orbital to be optimized."

"The method was applied to the simple, but non-trivial, LiH molecule in a periodic system, and we were able to obtain good agreement between the total energies from aperioidic and periodic plane-wave FCI calculations."

"Subsequent calculations showed that the correlation energy converged steadily as more virtual orbitals were included in the calculation."

"With 18 virtual orbitals the correlation energies were found to be converged to less than 1 kcal/mol."

"That the energies obtained using the H1–1 quantum computer were able to reproduce the FCI values to less than 11 milliHartree (6.9 kcal/mol) with a modest number of 500 shots performed; slightly less when corrected for noise."

Classical and Quantum Compression for Edge Computing: The Ubiquitous Data Dimensionality Reduction [69]

This is a machine-generated summary of:

Bagherian, Maryam; Chehade, Sarah; Whitney, Ben; Passian, Ali: Classical and quantum compression for edge computing: the ubiquitous data dimensionality reduction [69].

Published in: Computing (2023).

Link to original: https://doi.org/10.1007/s00607-023-01154-0

If you want to cite the papers, please refer to the original.

For technical reasons we could not place the page where the original quote is coming from.

Abstract-Summary

"Edge computing aims to address the challenges associated with communicating and transferring large amounts of data generated remotely to a data center in a timely and efficient manner."

"A survey of data compression algorithms with a focus on edge computing."

"Not all compression algorithms can accommodate the data type heterogeneity, tight processing and communication time constraints, or energy efficiency requirement characteristics of edge computing."

"We discuss specific examples of compression algorithms that are being explored in the context of edge computing."

Introduction

"Identification, as well as the development of data compression algorithms for EC devices is important for the implementation of EC."

"Given the heterogeneity of EC data types, volumes, processing, and communication time sensitivities, and resource constraint severities, it stands to reason to employ or devise compression algorithms that are tailored to such highly dynamic EC environments."

"We review existing data compression algorithms with an emphasis on their suitability for EC."

"We discuss the characteristics of a given data compression technique that may render it more suitable for EC than other techniques."

"With recent efforts to make use of quantum computing, "new" data types are being created, which may require quantum compression algorithms."

"We end with the quantum perspective of information theory with an introduction and a survey of quantum algorithms to compress quantum data."

Data at the Edge

"This section concerns recent work on compression algorithms designed for data at the edge."

"Before discussing data compression in the context of edge computing, we give a brief overview of the function of compression algorithms and the reasons that users may turn to them."

"If holding the entirety of the decompressed data in memory is impossible in this scenario, the user will require an algorithm whose output supports partial decompression or even computations performed directly on the compressed representation."

"Much of the data encountered on the edge is of a structured type (image data, for example) that has been previously studied and its compression algorithms are already available."

"One emergent theme in the literature is the design of new compression algorithms to handle novel edge data types."

"Beyond new compression algorithms, successful management of edge data requires an understanding of the relative strengths and weaknesses of existing, competing methods."

Compression at the Edge

"What makes data compression at the edge different than compression applied to data elsewhere on the network is the characteristic features of edge computing applications."

"At the edge, data compression improves not only the storage and latency but also the computation offloading."

"In the context of fog computing, such an optimization problem was recently reported in the work by Nguyen et al. [70], where studies with compression, first only at the mobile edge sensor, and then, at both the sensor and the fog server, showed that high levels of reduction in energy and delay cost can be obtained."

"Efficient data compression at the edge with minimized loss of information is a key to reducing energy consumption and enabling edge applications."

"Deepu et al. [71], aimed to reduce power consumption for IoT devices and investigated a data transmission method based on lossy and lossless data compression for wireless sensors."

Classical Compression Algorithms

"The original input data is approximated by Chebyshev polynomials, achieving very high compression ratios on serial data streams with minimal loss of scientific information."

"Power-efficient acquisition, processing, and storage, and time-efficient transmission of healthcare data at the edge is an active area in which enabling compression techniques are needed."

"The discrete transform, used in data compression, returns a vector of the same length as the input vector."

"With the steady increase in the processing power, Karhunen–Loeve transform (KLT) has been gaining increasing attention in signal and data processing for denoising and compression."

"For cases when the data matrix S is sparse (i.e., not dense) or can be represented in a domain in which the signal matrix has many zero or near zero components, compressed sensing can use the information about the sparsity in S to reconstruct the signal with a smaller sampling rate than permissible by (but without violating) the sampling theorem."

Quantum Compression Algorithms

"By invoking quantum computing at the edge where better sensors that can operate faster and generate more complete datasets, new data compression techniques and algorithms are necessary."

"The notion of an edge quantum computing device was recently introduced and described in the form of a realization of quantum sensors that generate quantum state data to be processed by quantum processors and quantum algorithms and communicated through quantum internet [72]."

"Several papers in the literature address this problem using different algorithms or schemes, quantum—classical data, quantum ensembles, state projection, etc. In

[73], a protocol for which an ensemble of qubits, storing all information about the quantum state, can in theory be perfectly compressed into exponentially fewer physical qubits is proposed."

"In [74], the authors consider the problem of One-Shot classical data compression while using quantum side information."

Conclusion

"The presented survey puts forth the ubiquity of data compression in the landscape of emerging IT technologies and IoT, as well as in relation to near-future visions of the quantum internet and the quantum IoT. It specifically emphasizes the need for efficient data compression for resource-constrained edge applications."

"As can be summarised from the presented material, the diversity in both the number of algorithms of potential for compression at the edge and in the edge computing use cases means that the unification of the pertinent concepts and approaches will be challenging."

"From a methodological point of view and future work directions, one may seek to explore new integral transforms of potential for data compression and examine their efficiency as the compression method of choice for edge computing."

"An interesting case could be that an index transform such as the Mehler–Fock transform, or Kontorovich–Lebedev transform which may find potential application in data compression."

Toward Implementing Efficient Image Processing Algorithms on Quantum Computers [75]

This is a machine-generated summary of:

Yan, Fei; Venegas-Andraca, Salvador E.; Hirota, Kaoru: Toward implementing efficient image processing algorithms on quantum computers [75].

Published in: Soft Computing (2022).

Link to original: https://doi.org/10.1007/s00500-021-06669-2

If you want to cite the papers, please refer to the original.

For technical reasons we could not place the page where the original quote is coming from.

Abstract-Summary

"Quantum information science is an interdisciplinary subject spanning physics, mathematics, and computer science."

"To promoting its mathematical and physical foundations, scientists and engineers have increasingly begun studying cross-disciplinary fields in quantum information processing, such as quantum machine learning, quantum neural networks, and quantum image processing (QIMP)."

"These issues include enhancing current models of quantum image representations, designing quantum algorithms for solving sophisticated operations, and developing physical equipment and software architecture for capturing and manipulating quantum images."

Introduction

"The expectation is that harnessing the properties of quantum-mechanical systems in QIMP (e.g., computational parallelism) will result in the realization of advanced technologies that will outperform, enhance, or complement existing and upcoming digital technologies for image and video processing tasks."

"QIMP has become a popular area of quantum research due to the ubiquity and primacy of digital image and video processing in modern life (Yan and Venegas-Andraca [76])."

"Different from all the previous works, the present paper seeks to arouse the interest of scientific and engineering communities toward the greater realization of QIMP-based technologies by identifying and discussing three primary issues using a few simulation experiments, the development of which will be most helpful for advancing the field of QIMP: improving current models of quantum-image representations, core level of designing quantum algorithms for solving sophisticated operations, and developing hardware (physical equipment) and software architectures (a new genre of blueprints) for capturing and manipulating quantum images."

Storage and Retrieval of Images in Quantum Systems

"We now analyze the FRQI representation model in order to quantify the amount of resources needed to store and retrieve information contained in quantum images."

"Current models of quantum image representation require extensive improvement to integrate and take full advantage of quantum-mechanical properties, as well as to design efficient strategies for image retrieval."

"Steps in QIMP research programs must include quantum-state tomography techniques (Banaszek et al. [77]) that, possibly combined with advanced computational paradigms such as machine learning (Youssry et al. [78]), would potentially lead to optimized quantum image retrieval processes [4]. A potential fruitful area for further development of QIMP is to test upcoming quantum image representation models, algorithms, and information retrieval techniques in fields in which images are likely to play a key role."

"A promising research avenue would be to design quantum image storage and retrieval methods suitable for use as input to algorithms from the emergent fields of quantum machine learning (Biamonte et al. [79]; Venegas-Andraca et al. [80]; Cruz-Santos et al. [81]), quantum computer vision (e.g., Yu et al. [82]), and quantum pattern recognition (Trugenberger [83])."

Algorithm Development and Algorithmic Speed-Up in QIMP

"We argue that imposing on QIMP the requirement of exponential speed-up as the main and/or only success criterion is a choice of limited scope that does not take into account the nature and goals of this discipline. (1) Exponential speed-up is the result of comparing efficient versus inefficient algorithms, and so achieving exponential speed-up on a quantum algorithm requires classical algorithms with exponential complexity as counterparts."

"Polynomial or exponential algorithmic speed-up must be a goal in QIMP, but, in addition to that aim, we should also consider other quality criteria, like the suitability of quantum images for human or machine consumption. (2) There are two approaches for estimating the complexity of an algorithm: asymptotic analysis and empirical analysis."

"Both classical and quantum versions of the Sobel operator algorithm are equivalent in terms of computational complexity."

The Road Ahead—A Proposal of Future Steps for QIMP

"To efficiently implement the retrieval process in QIMP, the measurement strategy for quantum images must be explored in depth (Yan et al. [84])."

"Although the development of QIMP technologies that are fully competitive with corresponding digital technologies is highly desirable, future research efforts must avoid attempting to realize quantum versions for every digital image processing algorithm, as not all digital image processing algorithms may be appropriate for implementation in the quantum computing realm."

"We must note that the development of physical equipment for capturing and processing quantum images is key for making QIMP as pervasive a field as digital image processing."

"Complete toolkits including loadable quantum modules and packages should be developed that can be employed by scientists and engineers as basic building blocks in designing hybrid quantum-classical image processing algorithms (Li et al. [85])."

Concluding Remarks

"Extending digital image processing to the quantum-computing realm, that is, QIMP, conjures similar expectations."

"QIMP has the potential to become a key component toward making quantum technology a pervasive field with huge impacts on many areas of science and technology, just like digital image processing is currently."

"To achieve this goal, the research agenda of QIMP must include the development of novel quantum image storage and retrieval techniques and to harness quantum phenomena (e.g., quantum entanglement) as a tool for image processing and analysis."

"Feynman's famous lecture title, "There is plenty of room at the bottom," has been an inspiration for many members of the QIMP community, including ourselves, to work toward the development of a branch of quantum science and engineering focused on storing, processing, and retrieving visual information using quantum systems, with the higher goal of contributing to the development of quantum-technology ecosystems."

Numerical Simulations of Noisy Quantum Circuits for Computational Chemistry [86]

This is a machine-generated summary of:

Wright, Jerimiah; Gowrishankar, Meenambika; Claudino, Daniel; Lotshaw, Phillip C.; Nguyen, Thien; McCaskey, Alexander J.; Humble, Travis S.: Numerical simulations of noisy quantum circuits for computational chemistry [86].

Published in: Materials Theory (2022).

Link to original: https://doi.org/10.1186/s41313-022-00047-7

If you want to cite the papers, please refer to the original.

For technical reasons we could not place the page where the original quote is coming from.

Abstract-Summary

"We investigate the behavior of these noisy quantum circuits using numerical simulations to estimate the accuracy and fidelity of the prepared quantum states relative to the ground truth obtained by conventional means."

"We implement several different types of ansatz circuits derived from unitary coupled cluster theory for the purposes of estimating the ground-state energy of sodium hydride using the variational quantum eigensolver algorithm."

Introduction

"By construction, the applied ansatz circuit has a best approximation to the true ground state of the given Hamiltonian operator, but the accuracy with which this state is found is not the leading metric for experimental validation."

"Claudino and others investigated the fidelity of several variational methods in approximating the ground state of few-electron molecular models [87]."

"Those results show that VQE methods with sophisticated ansatz circuits could approximate the ground electronic state with very high fidelity in the absence of circuit noise, depending on the ansatz and method of parameter selection."

"We investigate variations in parameter optimization due to noisy gate operations, testing the COBYLA [88] (derivative-free) and BFGS [89, 90] (gradient-based) optimizers, and in all cases we consider the influence of gate noise on the estimated energy and prepared state fidelity."

Variational Quantum Eigensolver Methods

"We use the qubit representation of the electronic Hamiltonian as the operator of interest, such that the estimated energy expectation value is where $|\psi(\theta)\rangle = U(\theta)|\psi(0)\rangle$ is a variable pure quantum state prepared by a (unitary) ansatz operator $U(\theta)$."

"The critical choice in applying the VQE method to a given Hamiltonian is selection of the ansatz operator $U(\theta)$ and the underlying reference state $\psi(0)$."

"The accuracy with which a given ansatz operator represents the true ground state of the Hamiltonian may be quantified using the fidelity where Ψ is the expected ground state in the qubit representation."

"For the four-qubit encoding of two-electrons described below, the resulting UCC doubles (UCCD) ansatz operator is reduced to a specific form. We also consider a singlet-adapted variation of the UCCSD ansatz for which the state is defined in a basis of restricted determinants."

"The above ansatz operators must be compiled into quantum circuits for execution."

Numerical Methods

"The compiled circuits are simulated numerically using the IBM aer simulator as the XACC backend, and we integrate a noise model that describes each gate by a noisy operation."

"In modeling noise in the quantum circuits, we assume a model by which each one- and two-qubit gates are followed by a depolarizing noise channel."

"The depolarizing noise model is a convenient device agnostic noise model that offers a coarse-grain representation for the loss of coherence caused by a noisy circuit, particularly for few-qubit numerical simulations [91]."

"Randomized compiling is known to yield a statistical model for the quantum circuit noise that is well approximated by the depolarizing model and motivates our use for the model here [92]."

"For our noise model, we use noise levels for the two-qubit CNOT gate that is always ten times the value of the single-qubit gate noise."

"We use the same noise level for all single-qubit gates."

Results

"There are noticeable differences in the parameters returned by COBYLA and L-BFGS with the latter showing a smoother variation across different levels of the noise parameter, especially at larger values of R. These differences, however, did not correspond to any significant difference to the reported energy or fidelity of the respective optimizer."

"Similar to the bare UCCD ansatz, the simulated ground state energy increases with increase in the noise parameter across all R. Again, we see a corresponding drop in fidelity as we increase the noise parameter, and minimal differences in the reported energy and fidelity when comparing the COBYLA and L-BFGS optimizers."

"The COBYLA optimizer has a greater variance than that of the L-BFGS optimizer, especially at higher values of R. When comparing the differences in optimal parameter selection between optimizers at the same level of noise, we report no significant difference to the resultant energy or fidelity between the respective optimizers."

Conclusions

"Across all ansatz circuits, we find that the noiseless simulations always recover the ideal energy and unit fidelity, while in the presence of noise, the depth of the ansatz circuit has the most important influence on the error in these quantities."

"Increases in gate depth of the ansatz necessarily increases the error in energy and lowers the fidelity and this is amplified at larger noise values."

"The differences observed between the bare UCCD and the randomly compiled UCCD circuits reflect differences in their single-qubit gate depth while the singlet-adapted UCCSD and ADAPT-VQE related ansatze, reflect differences in the number of two-qubit gates."

"As shown here, variability in the ansatz circuit depth due to increasing depolarizing noise leads to fluctuations in the energy and fidelity that are not observed within the fixed ansatz."

Own the Unknown: An Anticipatory Approach to Prepare Society for the Quantum age [93]

This is a machine-generated summary of:

de Jong, Eline: Own the Unknown: An Anticipatory Approach to Prepare Society for the Quantum Age [93].

Published in: Digital Society (2022).

Link to original: https://doi.org/10.1007/s44206-022-00020-4

If you want to cite the papers, please refer to the original.

For technical reasons we could not place the page where the original quote is coming from.

Abstract-Summary

"Advancing our capacities to acquire, process and transmit information, quantum technology has the potential to impact nearly all domains in society."

"The Netherlands Scientific Council for Government Policy (WRR) developed a framework for embedding so-called system technologies into society based on a historical analysis of how society dealt with such technologies in the past."

"The conceptual framework of system technologies is reinterpreted as an anticipatory strategy to prepare society for quantum technology and vice versa."

"By actively engaging in these processes, society enables itself to guide the development of quantum technology and its impact within society."

Introduction

"Asking any expert whether there is reason to be excited about quantum technology, and the answer will be unequivocal yes."

"I propose an anticipatory strategy to prepare society for quantum technology and vice versa."

"I then reflect on the question of what we can expect from quantum technology in the coming years and the challenge of preparing society for an unknown future."

"I introduce the concept of 'system technologies' (WRR, [94]) and use it to characterise quantum technologies as a technological family with a potentially broad and diverse impact on society."

"By elaborating on the dimensions of demystification (i), contextualisation (ii), engagement (iii), regulation (iv) and positioning (v), I lay down a roadmap for preparing society for quantum technology and guiding its impact."

"The article seeks to contribute to the nascent interdisciplinary discussion on the implications of quantum technology."

Why we Need to Start Talking about Quantum Technology Now

"The second class of 'quantum information technologies' includes subfields like quantum computation [95], quantum simulation and quantum communication."

"In the case of QIT, the reasoning that underlies this misconception is that the potential mathematical power of quantum technology is thought of as superior to that of classical computers, hence surpassing their capacities to solve mathematical problems across the board."

"To briefly look at some of these applications, it is useful to differentiate between three kinds of quantum technology: (1) quantum sensing, using the sensitivity of quantum systems to acquire information; (2) quantum computing, exploiting quantum properties to process information and perform computational tasks; and (3) quantum communication, harnessing quantum principles to transmit information."

"Quantum computing in the sense of using controlled quantum systems to solve more abstract problems—which is often meant when talking about 'quantum computers'—is technologically much more complex than simulation."

Characterising Quantum Technology as a 'System Technology'

"I conclude by reframing the five tasks that the WRR identified for embedding system technologies into society, as five dimensions of an anticipatory strategy for quantum technology."

"In order to shift the attention away from the technology itself and its economic implications towards the complex co-development of technology and society, the WRR introduced the new concept of 'system technology' (WRR, [94]: 127–128)."

"According to the WRR, history shows some general patterns in the introduction of system technologies into society (WRR, [94]: 134–136)."

"This provides us with five possible strategies to anticipate quantum technology and its implications for society by (1) demystifying the technology, (2) investing in its socio-technical context, (3) engaging other disciplines and stakeholders, (4) preparing regulation and (5) taking a position vis-à-vis international players and practices."

"The WRR firmly states that history unequivocally shows that regulation plays an important role when it comes to public acceptance and wide use of technologies."

Anticipatory Strategies for Guiding Quantum Technology

"An engaged discussion about how we want to use quantum technology in society from the start is key, as this could prevent technologies like quantum sensing from developing in such a way that it primarily becomes a tool for the powerful."

"As the defence industry is an important driver behind the development of quantum technology, progress in the field has implications for the warfare domain and hence, for national security (Krelina, [96])."

"Further developments in the field of quantum technology will highlight the importance of international collaboration."

"States should be aware of the strategic implications of quantum technology (Hoofnagle & Garfinkel, [97]; Der Derian & Wendt, [98]) and should therefore start with what can be called 'quantum diplomacy': Actively managing international relations with other states and international organisations with regard to the development, trade, and international regulation of quantum technology."

Conclusion and Discussion

"In order to anticipate the future role of quantum technology and the ethical, legal, social and policy implications that come with it, we should start to prepare society for this new system technology."

"I propose an anticipatory strategy to prepare society for quantum technology and vice versa."

"The proposed anticipatory strategy has five dimensions with specific challenges to address in a continual process of shaping technology and society in tandem: (1) countering unrealistic perceptions (demystification), (2) investing in a facilitating socio-technical environment (contextualisation), (3) engaging stakeholders and civil society (engagement), (4) creating flexible frameworks (regulation) (5) and developing international 'quantum diplomacy' (positioning)."

"This roadmap is exploratory and should be taken as a starting point for a discussion about the structural and long-term processes that will guide the co-evolution of quantum technology and society."

A Report on Teaching a Series of Online Lectures on Quantum Computing from CERN [99]

This is a machine-generated summary of:

Combarro, Elías F.; Vallecorsa, Sofia; Rodríguez-Muñiz, Luis J.; Aguilar-González, Álvaro; Ranilla, José; Di Meglio, Alberto: A report on teaching a series of online lectures on quantum computing from CERN [99].

Published in: The Journal of Supercomputing (2021).

Link to original: https://doi.org/10.1007/s11227-021-03847-9

If you want to cite the papers, please refer to the original.

For technical reasons we could not place the page where the original quote is coming from.

Abstract-Summary

"Quantum computing (QC) is one of the most promising new technologies for High Performance Computing."

"Its potential use in High Energy Physics has lead CERN, one of the top world users of large-scale distributed computing, to start programmes such as the Quantum Technology Initiative (QTI) to further assess and explore the applications of QC."

"As a part of QTI, CERN offered, in November–December 2020, a free, online series of lectures on quantum computing."

"Our lectures were designed with the objective of reducing the prerequisites to the bare minimum as well as focusing on hands-on, practical aspects of programming quantum computers and not on the mathematical analysis of the algorithms."

Introduction

"These hardware improvements have gone hand-in-hand with the proposal of new quantum algorithms which can be used for High Energy Physics (HEP) computational tasks and with the development of initial prototypes for concrete problems in the field [100–102]."

"At CERN, the research activities around quantum information processing technologies crystallized in 2020 in the creation of the Quantum Technology Initiative

(QTI), which encompasses the areas of quantum computing, quantum communication, quantum sensing and quantum simulation."

"As a part of such educational effort, from November the 6th until December the 18th 2020, CERN offered a weekly series of online lectures called "A Practical Introduction to Quantum Computing: From Qubits to Quantum Machine Learning and Beyond"."

"We report on the experience of designing and running such lectures in the broader context of CERN's quantum computing activities and of HPC education and training in general."

CERN and Large-Scale Computing Infrastructures from the Grid to Quantum

"30 years, CERN has actively contributed to evolving the international computing and data sharing infrastructures to support the computational needs of the High-Energy Physics community, but also for the more general benefit of broad scientific research."

"The interest in applications of machine learning and deep learning techniques for simulation, track reconstruction, and data classification has triggered an intense activity of research in accelerated computing architectures (GPUs, FPGAs, IPUs, TPUs, etc) and HPC infrastructures through several initiatives, including the CERN openlab, started in 2001 with the mandate of setting up R&D projects in support of the physics research programmes."

"In 2020, CERN openlab promoted the creation of the Quantum Technology Initiative (QTI) to explore the possibilities of the main four areas of today's quantum technologies, computing, sensing and metrology, communications, and quantum simulation and information processing, for the High Energy Physics community in general and for CERN in particular."

Teaching Quantum Computing: Challenges and Opportunities

"All this makes the task of teaching quantum computing a challenging one, especially when the prospective students have backgrounds which cover only part of the concepts required to describe and study quantum algorithms (for instance, if they come only from the field of Computer Science or Software Engineering)."

"The authors of [103] mention that there is still a large number of challenges in developing effective pedagogy to train students who are familiar with classical algorithms in the different paradigm associated with quantum algorithms and suggest that one thing that a higher-education institution could do is introduce an intro-level quantum course focusing on either the hardware or algorithms aspects of quantum information science."

"Similar recommendations are given in [104–108], which also report successful experiences on teaching introductory QC courses to students with different backgrounds (including Computer Science and Software Engineering majors), with a hands-on approach and a focus on implementing quantum algorithms on quantum simulators and actual quantum computers."

A Practical Introduction to Quantum Computing: CERN'S Online Lectures on Quantum Computing

"CERN's introductory lectures on quantum computing were conceived, from the very beginning, to be accessible for an audience as wide as possible."

"Throughout all the lectures, we adopted an axiomatic approach, describing the elements of quantum computing (qubits, unitary transformations and measurements) as if they were abstract data types and operations and without explaining their physical implementation."

"These two directives were clearly stated in the title of the lectures, A Practical Introduction to Quantum Computing, with the "practical" and "introduction" aspects corresponding to the second and first guidelines, respectively."

"The series of lectures was designed to be an introduction to quantum computing that could be followed by any person with only knowledge of basic linear algebra."

"One difference of our lectures with other approaches to teaching quantum computing at the introductory level is that we expended a large fraction of the time explaining concepts and examples which use just one or two qubits."

Evaluation of the Lectures: Reception, Survey and Results

"We prepared a survey with questions about the lectures and about the way the participants use computational resources in their work, research and studies."

"Regarding the current occupation, participants from universities scored significantly higher than the rest the teaching materials and the ease of use/access of the lectures."

"As for the computer science techniques and tools currently used, participants using Artificial Intelligence scored significantly lower than the rest the contents of the lectures."

"One is that participants using cloud computing as their computing environment scored significantly higher both the approach/methodology and the contents of the lectures."

"Another one that participants using quantum simulators as computing environment scored significantly higher than the rest in three questions: contents, teaching materials, ease of use/access."

"Participants considering QC useful for their research scored significantly higher than the rest the possibility of asking questions."

Conclusions

"This fact endorses that the approach, not focused on mathematics and physics but on algorithmic and implementation elements, is a useful methodology, with great acceptance by the participants, who repeatedly mentioned this aspect on their answers to a satisfaction survey."

"The most important result from the survey is that the perceived knowledge about QC increased very significantly and all the aspects about methodology, contents and materials were very highly scored."

"Another remarkable conclusion is that, even with a limited number of teaching hours, the approach of focusing on algorithmic and programming aspects of quantum computing, paying relatively less attention to the mathematical formalism and physical implementations, allowed us to cover topics that are considered "advanced" and not usually taught in introductory QC courses, such as variational algorithms, quantum annealing and quantum machine learning."

"For the future, it would be interesting to study the possibility of exporting some of the methodological aspects used in these lectures to other QC teaching initiatives."

Quantum Computing Opportunities in Renewable Energy [109]

This is a machine-generated summary of:

Giani, Annarita; Eldredge, Zachary: Quantum Computing Opportunities in Renewable Energy [109].

Published in: SN Computer Science (2021).

Link to original: https://doi.org/10.1007/s42979-021-00786-3

If you want to cite the papers, please refer to the original.

For technical reasons we could not place the page where the original quote is coming from.

Abstract-Summary

"Quantum computers are capable of using problem-solving approaches which are not available to classical computers."

"There is a need to identify good near-term problems to demonstrate quantum computing's problem-solving potential."

"We identify a few places where quantum computing is most likely to contribute to renewable energy problems: in simulation, in scheduling and dispatch, and in reliability analyses."

"The problems have the common theme that there are potential future issues concerning scalability of current approaches that quantum computing may address."

"We then recommend potentially fruitful areas of crossover research to advance applications of quantum computing and renewable energy."

Introduction

"Quantum computers are capable of using problem-solving approaches which are not available to classical computers."

"The most interesting quantum approaches are those which allow larger problems to be solved without dramatically increasing the time required for computation."

"There is a need to identify good near-term problems, which are still small and require special-formulation, to demonstrate quantum computing's problem-solving potential."

"We identify several opportunities for the emerging field of quantum computing to contribute to the renewable energy transition."

"Having identified these subject areas for potential collaboration, we make suggestions for the directions and types of research programs needed in each areas: Crossover research between quantum computing and operations research which can translate between the most promising approaches in quantum mechanics and useful problems."

"Experimental realization of basic quantum computing implementations of renewable-energy-relevant problems to identify promising approaches even in areas where exact theoretical knowledge about quantum performance is still inaccessible."

Background

"The next subsections present brief overviews of the current state of the art and challenges in variable renewable energy and quantum computing."

"Qubits in a quantum computer do not need to fall into the 0/1 binary of classical bits, but can occupy interesting and useful superposition states in the course of the computation."

"Quantum computers are able to leverage the unique aspects of quantum mechanics discussed earlier—superposition, interference, and entanglement—to perform algorithms that a classical machine cannot."

"To D-Wave's superconducting-based technology, other technologies such as nonlinear optical resonators have also been used to create" analog" quantum computers performing optimization problems [110]."

"Harrow–Hassidim–Lloyd (HHL): an algorithm for extracting information from the solutions to linear systems, reducing the time required on a system of N equations from scaling with N to log N, part of a family of techniques which address linear systems problems on quantum computers."

Areas of Possible Application in Energy Systems

"Quantum computers are well suited to simulate the Schrödinger equation that governs the behavior of chemistry-scale problems."

"Early efforts to capture chemical behavior have included the use of Variational Quantum Eigensolvers (VQE), a hybrid quantum/classical algorithm which successfully computed some test problems, such as the electronic states of a water molecule [111]."

"Current Challenges and Description of Quantum Approach The efficacy of quantum approaches in forecasting depends on identifying key places where quantum computers would accelerate or improve the performance of difficult aspects of the computational problem."

"Two approaches to these complicated optimization problems using quantum computers should be examined: quantum annealing and quantum approximate optimization algorithms (QAOA)."

"Current Challenges and Description of Quantum Approach Quantum computer scientists have found several algorithms which may be of use to the electrical power engineer interested in system stability and reliability."

Current Investment in Quantum Computing

"The scale of investment in quantum computing is large and growing, with money flowing in from industrial and government sources."

"The private sector has also made significant investment into quantum computing."

"In 2019, Google announced that a quantum computer based on superconducting qubits had achieved a milestone known as" quantum supremacy", meaning a computation was performed on a quantum computer that would have been infeasible for a classical one."

"IBM, Intel, and Google have focused on constructing general-purpose quantum computers using superconducting junctions as building blocks."

Conclusion and Recommendations

"We have attempted to identify important areas in the field of renewable energy that we believe quantum computers may be able to contribute to as the hardware develops."

"Further research is clearly needed to determine the scope and scale of potential improvements in renewable energy computing enabled by quantum technologies."

"Research projects in this area could also work to clarify current limitations in optimization and establish quantitative benchmarks for quantum computing."

"Work is needed to characterize the performance of quantum computers in small optimization problems with operational relevance."

"By continuing to work together, energy researchers and quantum computer scientists will ensure that when operational quantum hardware becomes available, the industry is prepared to deploy it for useful problems."

"Some of the transformative technologies needed to create a renewable energy system may, in fact, be the quantum computing technologies that will be developed in the next decade."

State-of-the-Art Survey of Quantum Cryptography [112]

This is a machine-generated summary of:

Kumar, Ajay; Garhwal, Sunita: State-of-the-Art Survey of Quantum Cryptography [112].

Published in: Archives of Computational Methods in Engineering (2021).

Link to original: https://doi.org/10.1007/s11831-021-09561-2

If you want to cite the papers, please refer to the original.

For technical reasons we could not place the page where the original quote is coming from.

Abstract-Summary

"With the significant development in the area of quantum computing, there is a need for unconditional security in confidential information."

"Quantum key distribution protocols are proven secure if all devices are perfect (in terms of technologies and proper protocol operations)."

"The purpose of this survey article is to carry out a systematic review in the area of quantum cryptography by covering various aspects of non-deterministic quantum key distribution protocols, quantum secure direct communication, semi-quantum key distribution, secure multiparty communication protocol, post-quantum cryptography and device-independent cryptography techniques."

"We also discussed various experimental work carried out in the area of quantum cryptography, various attacks and challenges relative to the paradigm shift from classical cryptography to quantum cryptography."

Introduction

"Once quantum computer will be available, Shor's algorithm will give security threats to all classical cryptographic protocol [113]."

"Compared with previous existing survey papers [114–119], our survey introduces the in-depth discussion of Quantum key distribution protocols, reviews the existing work published up to 2020, serving as a guide for other researchers to understand and apply the existing protocols, current research directions and discusses several open problems."

"For a better understanding of the state-of-the-art in quantum cryptography, we surveyed with the following goals: We review various concepts and terminologies used for understanding quantum protocols."

"An exhaustive survey on deterministic protocols for quantum secure communication without the shared secret key."

"To identify and discusses the current trends in quantum cryptography like satellite-based communication, device-independent cryptography and high-dimensional Quantum key distribution."

"We survey the existing literature on semi-quantum key distribution protocols."

Preliminaries

"Robustness: The legitimate user (Alice/Bob) will detect errors if Eve attempt to obtain or alter the information No Cloning Theorem: [120] It states that an unknown quantum state cannot be cloned."

"Quantum Entanglement: State of two or more quantum particles are entangled if many of the physical properties of the particles are strongly correlated."

"The heart of quantum cryptography is entangled states."

"Quantum one-time pad allows Alice and bob to share the secret message (in the form of private quantum states) over a public quantum channel."

"Brandao and Oppenheim [121] carried out the work on the quantum one-time pad for sharing quantum messages by considering that Alice and Bob's state is related with Eve state."

"Privacy Amplification: In the quantum protocol, privacy amplification is performed to reduce the amount of information known to Eve by shrinking the key."

Research Methodology

"Paper selection consists of following two phases:"

"Title and Abstract Level Screening: Initially, We had selected papers from 1964 to 2020, and two essential papers of 1927 and 1935 are considered."

"Full-Text Screening: In this phase, we had analyzed the papers based on the full text."

"If two or more papers were contributed by the same authors and their significant contribution is same, we considered the most relevant paper with a significant contribution."

Discrete and Continuous Variable Quantum Key Distribution Protocol

"Quantum key distribution protocols are classified into Discrete variable QKD and Continuous variable QKD protocol."

"Protocol based on continuous variable offer advantage over discrete because coherent light with photon can easily producible using laser than single-photon [122]."

"Garcia-Patron and Cerf [123] proposed a continuous-variable QKD protocol based on squeezed states and heterodyne detection for obtaining higher security key rate over the noisy line."

"Cerf and Grangier [124] surveyed various continuous-variable Quantum key distribution protocol."

"Leverrier and Grangier [125] proposed two continuous variable QKD protocols with discrete modulation using two and four coherent states."

"Papanastasiou and Pirandola [126] designed continuous-variable QKD protocol using discrete-alphabet encoding."

Quantum Attacks

"Eve's attacks can be classified into individual, collective and coherent attacks."

"Individual Attack: Eve prepares each ancilla qubit independently, interact with each qubit on quantum channel independently and measure independently."

"Collective Attack: Eve prepares each ancilla qubit independently, interact with each qubit on quantum channel independently and measure jointly all ancilla qubits."

"Coherent Attack (Joint Attack): Eve prepares entangled states of the ancilla qubits, interact with qubits on the channel and then measure all ancilla qubits collectively."

"In timing side-channel attack, timing information disclosed by Communicating parties (Alice and Bob) during the public discussion is used by Eve to access the significant part of the secret key."

"Qi and others [127] introduced the time-shift attack in which Eve shift the arrival time of signal pulse or synchronization pulses or both between Alice and Bob."

"Various quantum attacks can be classified into attack at source (Photon number splitting attack [128–130], Phase remapping attack [131, 132], Laser Seeding (Sun and others [133] etc) and attack at detection (Timing-side channel attack [134], Faked state attack [135], Time-shift attack [127, 136] and Polarization shift [137])."

Quantum Key Distribution Protocol

"In prepare and measure protocol, one party (say Alice) prepares a quantum state and sends the prepared quantum state to another party (say Bob), who will measure it."

"In BB84, Alice communicate the basis in which she prepared her qubits on a classical authenticate channel to Bob."

"Garcia-Patron and others [138] proposed single-photon two-qubit quantum logic for simulating the optimal individual attack on BB84 protocol without quantum memory."

"Boyer and others [139] proposed a protocol BB84-INFO-z (Identical to BB84, except information bits are in z-basis) and found that the modification in BB84 does not harm its security against collective attacks."

"There are 1/3 chances that Alice and Bob measure in compatible bases (E91 original protocol consider three bases)."

"Disadvantage of Six-State Protocol: In the six-state protocol, Bob has a quantum memory, and he performs all its measurement after Alice reveals the basis."

"In SARG04 protocol, Alice never announces her basis to Bob."

Quantum Secure Direct Communication (QSDC)

"This protocol is called ping-pong as the travelling photon travels from Bob to Alice and back to Bob."

"In their proposed protocol, Alice prepares n-pairs of maximally entangled state and send half of the qubits to Bob."

"Following steps are used in Chamoli and Bhandari's ping-pong protocol [140]: Alice prepares initial state of three photons in one of the eight GHZ states."

"In control mode (Similar as in the original ping-pong protocol), Bob and Charlie perform measurement in z-basis and inform their results to Alice through a public channel."

"Sarvaghad-Moghaddam [141] proposed an efficient and secure protocol using the concept of entanglement swapping for bidirectional quantum secure direct communication under the controller permission."

"Tan and Cai [142] pointed out that in quantum dialogue protocols, half of the message between Alice and Bob is leaked through classical public communication."

Semi-Quantum Key Distribution Protocol

"Further, Krawec [143] designed the Mediated semi-quantum key distribution protocol (multi-user quantum key distribution protocol) using Bell basis for allowing two classical or limited semi-quantum users (Alice and Bob) to establish a secret key using the untrusted full quantum server/center."

"Yu and others [144] designed the first SQKD protocol free from all attack and without using authentic classical channels known as Authenticated semi-quantum key distribution (ASQKD)."

"Zou and others [145] proposed a semi-quantum protocol without involving the classical Alice's measurement capability."

"Tsai and others [146] proposed a semi-quantum secret sharing protocol using W-state for three parties and found that the protocol is free from the well-known attacks."

"Lin and others [147] proposed a semi-quantum protocol to share a secret key between two classical users with the help of third untrusted party."

"Yan and others [148] proposed a semi-quantum protocol to transmit a secret message between classical Bob and quantum Alice using Bell states."

Secure Multiparty Communication (SMPC)

"Sun and others [149] proposed fairness (No one alone is not able to determine the key) multiparty quantum key protocol using maximally entangled six-qubit states."

"Sun and others [150] proposed a single qubit state protocol for multiparty quantum key agreement by performing an exclusive-OR operation on all the parties without the explicit need of entanglement states, joint measurement and unitary operations."

"Huang and others [151] proposed an efficient, fair and secure multiparty quantum key agreement protocol using single photons in travelling mode."

"Cao and others [152] proposed a multi-party quantum key agreement protocol for travelling mode based on non-orthogonal quantum pairs, Bell states and their dualities by mixed dense encoding."

"He and others [153] proposed a high-efficiency three-party quantum key agreement protocol by utilizing two-photon polarization entangled Bell states and a few single-photon polarization states."

Device Independent Cryptography

"Tomamichel and others [154] showed that the standard BB84 QKD scheme is one-sided device-independent QKD by considering Bob's quantum apparatus as malicious, and Alice apparatus is a trusted one."

"Lo and others [155] introduced the concept of measurement device-independent QKD for removing all detector side channels attacks."

"Tang and others [156] performed the first experimental realization of measurement device-independent QKD by considering the state preparation flaws and distributed secure keys up to 40 KM."

"Semi-device-independent QKD provides secure key distribution for one way prepare and measure protocols [157]."

"Detector Device-Independent Quantum Cryptography: To overcome the limitations of measurement-device independent QKD (Security key rate and Interface of two photons), Lim and others [158] and Gonzalez [159] proposed the concept of detector device-independent quantum cryptography the combine the security of measurement-device independent quantum cryptography with the efficiency of conventional QKD."

"Sajeed and others [160] demonstrated that detector-device-independent QKD is not secure against side-channel attacks."

Post Quantum Cryptography

"Post-quantum algorithms deal with cryptosystem that runs on a conventional computer but secure against attacks by quantum computer [161]."

"Post-quantum cryptography schemes are classified into code-based cryptography, Lattice-based Cryptography, Hash-based Cryptography, Multivariate-quadratic equations cryptography."

"The main issue of a code-based cryptosystem is the key size (megabyte) for higher security."

"Although Researchers had proposed few code-based cryptography schemes; attacks have been proposed corresponding to these schemes."

"There is a possibility of new code-based cryptography approach to be proposed that remain secure with the quantum attack."

"Lattice-Based Cryptography: Hoffstein and others [162] introduced NTRU public cryptosystem with a smaller key size than McEliece cryptosystem."

"To gain confidence against quantum attack, more research is needed to be carried out on lattice-based cryptography."

"Dods and others [163] and Hulsing [164] proposed the improved hash-based cryptography schemes using better one-time signatures to decrease the signature size."

Latest Trends and Concluding Remarks

"Amer and others [165] proposed a semi-quantum key distribution protocol for tolerating high-level of noise by considering the advantage of a two-way quantum channel."

"Arrighi and Salvail [166] introduced the concept of quantum blind computation and proposed a protocol for carrying out the blind quantum computation."

"Broadbent and others [167] proposed a protocol for blind quantum computation."

"High-dimensional Quantum Key Distribution is an efficient and robust way to encode information with higher key rate."

"Qi and Siopsis [168] studied the performance of position-based quantum cryptography protocols over a noisy channel by assuming that no entanglement is pre-shared between adversaries."

"The Quantum Technologies Group of the University of Geneva, ID Quantique and Corning Incorporated performed a successful Quantum key distribution at a distance of 421 KM using a three-state time-bin protocol with decoy approach and 2.5 GHZ repetition rate [169]."

Quantum Simulations of Materials on near-Term Quantum Computers [170]

This is a machine-generated summary of:

Ma, He; Govoni, Marco; Galli, Giulia: Quantum simulations of materials on near-term quantum computers [170].

Published in: npj Computational Materials (2020).

Link to original: https://doi.org/10.1038/s41524-020-00353-z

If you want to cite the papers, please refer to the original.

For technical reasons we could not place the page where the original quote is coming from.

Abstract-Summary

"Quantum computers hold promise to enable efficient simulations of the properties of molecules and materials; however, at present they only permit ab initio calculations of a few atoms, due to a limited number of qubits."

"In order to harness the power of near-term quantum computers for simulations of larger systems, it is desirable to develop hybrid quantum-classical methods where the quantum computation is restricted to a small portion of the system."

"We perform calculations on quantum computers and show that they yield results in agreement with those obtained with exact diagonalization on classical architectures, paving the way to simulations of realistic materials on near-term quantum computers."

Introduction

"Several theoretical and computational methods have been developed over the years to treat systems exhibiting strongly-correlated electronic states, including dynamical mean-field theory [171, 172] and quantum Monte-Carlo [173, 174]; in addition, ab initio quantum chemistry methods, traditionally developed for molecules, have been recently applied to solid state problems as well [175]."

"Quantum computers hold promise to enable efficient quantum mechanical simulations of weakly and strongly-correlated molecules and materials alike [176–189]; in particular when using quantum computers, one is able to simulate systems of interacting electrons exponentially faster than using classical computers."

"These spin-defects are promising platforms for solid-state quantum information technologies, and they exhibit strongly-correlated electronic states that are critical for the initialization and read-out of their spin states [190–195]."

"Based on the effective Hamiltonian derived from the quantum embedding theory, we investigated the strongly-correlated electronic states of the NV center in diamond using quantum phase estimation algorithm (PEA) [176, 196–198] and variational quantum eigensolvers (VQE) [199–201], and we show that quantum simulations yield results in agreement with those obtained with classical full configuration interaction (FCI) calculations."

Results

"Starting from an atomistic structural model of materials (e.g., obtained from DFT calculations or molecular dynamics simulations), we identify active regions with strongly-correlated electrons, which we describe with an effective Hamiltonian that includes the effect of the environment on the active region."

"We apply the theory to spin-defects including NV and SiV in diamond [202, 203] and Cr in 4H-SiC [204, 205]. Most of these defects' excited states are strongly-correlated (they cannot be represented by a single Slater determinant of single-particle orbitals), as shown e.g., for the NV center [206] in diamond by Bockstedte and others [207] using cRPA calculations."

"Upon solving the effective Hamiltonian, we predict the lowest excited state to be a ^1E state arising from e $\rightarrow$ e spin-flip transition, with excitation energy of 1.09 (0.86) eV based on embedding calculations beyond (within) the RPA."

Discussion

"With the goal of providing a strategy to solve complex materials problems on NISQ computers, we proposed a first-principles quantum embedding theory where appropriate active regions of a material and their environment are described with different levels of accuracy, and the whole system is treated quantum mechanically."

"Our method overcomes the commonly used random phase approximation, which neglects exchange-correlation effects; importantly it is applicable to heterogeneous materials and scalable to large systems, due to the algorithms used here to compute response functions [208, 209]."

"For systems where the electronic structure of the active region is expected to influence that of the host material, a self-consistent cycle in the calculation of the screened Coulomb interaction of the environment can be easily added to the approach."

"We expect the strategy presented here to be widely applicable to carry out quantum simulations of materials on near-term quantum computers."

Methods

"All ground state DFT calculations are performed with the Quantum Espresso code [210] using the plane-wave pseudopotential formalism."

"All geometries are relaxed with spin-unrestricted DFT calculations using the Perdew-Burke-Ernzerhof (PBE) functional [211] until forces acting on atoms are smaller than 0.013 eV / Å. NV and SiV in diamond are modeled with 216-atom supercells; Cr in 4H-SiC is modeled with a 128-atom supercell."

"Construction of effective Hamiltonians is performed with the WEST code [212], starting from wavefunctions of spin-restricted DFT calculations."

"In PEA simulations, the Jordan–Wigner transformation [213] is used to map the fermionic effective Hamiltonian to a qubit Hamiltonian, and Pauli operators with prefactors smaller than 10^{-6} a.u."

"For the simulation of the $M_S = 1$ state, the resulting qubit Hamiltonian acts on four qubits and there are two variational parameters in the UCCSD ansatz."

"For the simulation of the $M_S = 0$ state, we fixed the occupation of the a orbital and the resulting qubit Hamiltonian acts on 2 qubits."

Engineering Long Spin Coherence Times of Spin–Orbit Qubits in Silicon [214]

This is a machine-generated summary of:

Kobayashi, Takashi; Salfi, Joseph; Chua, Cassandra; van der Heijden, Joost; House, Matthew G.; Culcer, Dimitrie; Hutchison, Wayne D.; Johnson, Brett C.;

McCallum, Jeff C.; Riemann, Helge; Abrosimov, Nikolay V.; Becker, Peter; Pohl, Hans-Joachim; Simmons, Michelle Y.; Rogge, Sven: Engineering long spin coherence times of spin–orbit qubits in silicon [214].

Published in: Nature Materials (2020).

Link to original: https://doi.org/10.1038/s41563-020-0743-3

If you want to cite the papers, please refer to the original.

For technical reasons we could not place the page where the original quote is coming from.

Abstract-Summary

"Electron-spin qubits have long coherence times suitable for quantum technologies."

"Spin–orbit coupling promises to greatly improve spin qubit scalability and functionality, allowing qubit coupling via photons, phonons or mutual capacitances, and enabling the realization of engineered hybrid and topological quantum systems."

"We demonstrate ultra-long coherence times of 10 ms for holes where spin–orbit coupling yields quantized total angular momentum."

"These results open a pathway to develop new artificial quantum systems and to improve the functionality and scalability of spin-based quantum technologies."

Main

"We show that by engineering the quadrupole degree of freedom associated with the linear response of generalized spins to environmental electric fields using strain, total angular momentum $J = 3/2$ spin–orbit systems can be protected against decoherence and have a T_2 that rivals the best results for spin $S = 1/2$ electrons in silicon [215–217] and $S = 1$ nitrogen-vacancy centres in diamond [218–222]."

"The generalized spin has compelling properties for building spin–orbit qubits that couple to electric or elastic fields while maintaining long T_2."

"The observed improvement of T_2 in the strained sample reflects a reduced sensitivity of the qubit to electric-field noise, which we now analyse for generalized-spin qubits in strained ^{28}Si:B."

"It has recently been shown that monolithically fabricated silicon [223] field-effect transistors can present a very small strain environment [224], and thus the engineering of intentional strain should enable the implementation of acceptor-based generalized-spin qubits with long T_2 values and spin–orbit functionality in field-effect transistors."

Methods

"Si:B holes couple to uniform magnetic, electric and elastic fields as described in the literature [225, 226]."

"The coupling Hamiltonian to the fields is expressed as follows:where the x, y and z axes correspond to the [100], [010] and [001] axes of the silicon crystal, respectively, μ_B is the Bohr magneton, E_i are electric fields and ε_{ij} are normal (shear) strains for $i = j$ ($i \neq j$)."

"Each sample is mounted to the cavity in an independent experimental run to avoid overlapping signals from different samples."

"To couple to the cavity modes, the samples are directly placed on the cavity surface and closely fitted so that the polished silicon surface faces the cavity structure, and then fixed by GE Varnish."

"For the relaxed (strained) sample, $\omega_{MW}/2\pi$ of 6.255 GHz (6.331 GHz) is used."

Materials Science for Quantum Information Science and Technology [227]

This is a machine-generated summary of:

Richardson, Christopher J. K.; Lordi, Vincenzo; Misra, Shashank; Shabani, Javad: Materials science for quantum information science and technology [227].

Published in: MRS Bulletin (2020).

Link to original: https://doi.org/10.1557/mrs.2020.147

If you want to cite the papers, please refer to the original.

For technical reasons we could not place the page where the original quote is coming from.

Abstract-Summary

"At the heart of these technologies is the use of a quantum object to contain information, called a quantum bit or qubit."

"Current realizations of qubits exist in a broad variety of material systems, including individual spins in semiconductors or insulators, superconducting circuits, and trapped ions."

"We discuss some of the needs and opportunities for contributions to advance the fundamental understanding of materials used in quantum information applications."

Introduction

"Rather, quantum information systems are being designed to extend the state of the art for tackling only the most difficult problems."

"The uniquely quantum principles of superposition and entanglement provide for an enormous increase in information density and computational parallelism but come at the cost of needing to develop quantum analogues for bit operations, and to build these elemental operations up into algorithms for computation, communication, and sensing."

"The desire to have precisely two addressable quantum states requires the use of systems that have non-degenerate energy levels that are well isolated."

"A current recurring theme in quantum systems, spanning computation, communication, and sensing applications, is that they demonstrate low fidelity for initialization, manipulation, and readout, or they do not scale to large enough numbers of qubits to be practically useful."

Sidebar

"Manipulability of the quantum states with high fidelity to implement a universal gate set."

"They are rooted in the need for proper qubits to encode and maintain quantum states, along with the ability to execute prescribed transformations (algorithms) on those states to produce, in the end, a determined, measurable computational result."

"The ability to initialize the qubits to known states before a computation occurs relates to the ability to control a computation so that the solution to a specific problem is found by the execution of the quantum program."

"The ability to transmit flying qubits between specified locations with little decoherence of the quantum state."

"While these additional criteria pertain specifically to quantum communications protocols that may require exchange of quantum coherent states and transmission of quantum information to third parties, they may also be considered optional criteria for quantum computing."

Qubits and Materials

"The small energy scales of many qubit realizations (e.g., superconducting circuits, spins in semiconductors), combined with the requirements of the DiVincenzo criteria, drive engineering choices to operate QIST devices at the lowest possible temperatures achievable by current technology."

"Quantum sensors are often desired to operate at room temperature, so it may be assumed that many quantum-sensing devices will leverage qubit technologies operable in ambient laboratory environments for the foreseeable future."

"[228] For computing, there have been large numbers of demonstrations of qubits in a large array of quantum mechanical systems and a plethora of materials systems."

"The state of experimental quantum computing has evolved beyond merely searching for a quantum mechanical system to demonstrate realizations of QIST principles, into advancing hardware implementations that dramatically improve qubit performance and scalability and demonstrations of useful quantum computations."

Materials Selection

"While there are indeed vestiges and continued research in designing workaround techniques to leverage a wider range of materials, the leading materials for computational qubits are entrenched, but not finalized."

"For spin qubits, nuclear spin-free ($I = 0$) host materials are mandatory for computation and highly desirable for sensing."

"Despite manipulation of the nuclear spin bath presented by the host material to minimize its impact [229], there has been a down-selection of optimal materials for QIST."

"Another major qubit technology is trapped ions, which has evolved toward presumably scalable surface trap geometries [230] that introduce a number of materials challenges associated with the choice of electrode material."

"There are no front runners in terms of the best dielectric, conductor, or optimal deposition techniques to use for surface traps, which offer broad opportunities for new materials to be inserted into qubit devices for both sensing and computing."

Understanding Noise from a Materials Perspective

"Without exploring the specific origin, qubits experience charge noise, which is spatial and temporal variation of the electrical potential, and flux noise, which is spatial and temporal variation of the magnetic vector potential."

"[231] However, perhaps more insidious are noise sources intrinsic to the qubits themselves, originating from the materials used in their fabrication."

"A plethora of possible materials defects and properties that may generate noise within the qubit and contribute to decoherence or unwanted cross-coupling among qubits or quantum states stored within them make optimization of qubit fabrication and materials choices challenging."

"Mitigating the materials origins of noise creates great opportunity for improving their intrinsic performance and enabling across-the-board improvements in the ability to leverage the quantum resources they represent, including enabling greater scalability and/or the ability to implement quantum error correction."

"Research into computational techniques that are beyond-DFT could provide valuable insights into the microscopic origins of noise in quantum information materials and devices."

Materials Synthesis and Device Fabrication

"The requirement of having a long-lived quantum state dictates that material losses need to be eliminated."

"The path toward this potential requirement creates a number of challenges in materials synthesis and device fabrication."

"For materials synthesis, challenges exist in the growth and integration of novel materials that have not been used in the conventional electronics industry."

"For starters, intrinsic losses can be circumvented through materials selection, while impurity and defect losses must be addressed through synthesis and fabrication."

"The atomic-scale details of materials interfaces and variations in short-range ordering become important."

"[232, 233] It is possible that new materials systems and epitaxial growth techniques will advance some aspects of QIST devices."

"[234] Many QIST devices are fabricated with electron-beam lithography or focused ion beams to achieve the features with critical dimensions below 100 nm."

Materials Characterization

"This is because the relatively low yield of devices in academic laboratories and overhead required to produce and test devices has resulted in the physics community relying on just those techniques and materials that work as determined by correlation between specific process recipes of available tools and qubit characterization."

"The fundamental contribution of materials science is needed because today's qubits operate with characteristic energies that are on the order of 30 μeV, and our knowledge of materials in this energy regime is sparse."

"Improving the current understanding of material properties and development of advanced materials with engineered properties at cryogenic temperatures are clear contributions that materials science research can make."

"[235–239] Investment is needed in understanding surface chemistry, passivation, and the connection between atomic configuration and composition to the electronic structure of the surrounding material region on the tens-of-nanometers length scale, with μeV or better energy resolution, and at cryogenic temperatures."

Emerging Engineered Materials

"Different quantum platforms have their own merits, and limitations and successful systems may include hybrid approaches, since it has become clear that no single platform possesses all the advantages, thus interaction between dissimilar qubits is necessary."

"One example of a near-term hybrid system is the adoption of circuit quantum electrodynamic approaches for spin qubits; another is the use of dispersive readout (a measurement technique used in superconducting qubits) in silicon germanium quantum dot qubits."

"[240, 241] Other types of quantum materials may enable a whole new class of qubit devices."

"It is possible that the technological impact of topological protected states is reduced if MZM qubits are not developed at the rate of or faster than other qubit systems, since traditional quantum error correction approaches using surface codes based on stabilizers on a two-dimensional array [242] and other logical qubit configurations can provide comparable benefits from a wider range of qubit implementations."

Critical Non-qubit Materials Technologies

"A number of auxiliary technologies for QIST can also benefit from materials advancements."

"Some examples include surface traps with integrated photonics for ion traps [243], 3D superconducting circuit integration, [244] superconducting ribbon cabling for both controlled thermal and electrical transport."

"[245, 246] Functional coatings and adhesives that can withstand dramatic differences in thermal expansion coefficients could benefit all qubits that operate at millikelvin temperatures."

"[247] Furthermore, the materials advancements for improved qubits can be leveraged for improved quantum sensors and communications devices, as outlined previously."

Outlook

"Quantum information science is an interdisciplinary technology, with materials science having a clear and important role to play."

"For solid-state qubits, materials synthesis and device physics will undoubtedly play a major role in advancing the state-of-the-art technology."

"Academic labs that focus on individual and pairs of qubits, emerging qubit materials, and fundamental correlations between materials and qubit performance will continue to provide value to the maturing QIST community by training its workforce and exploring innovations that may lead to disruptive technologies."

"It is not clear what materials and design will move QIST beyond NISQ, nor is it clear how long this will take, but materials science will continue to be a critical aspect of this technology moving forward."

"As centers focused on quantum information science and technologies emerge, it is important that materials scientists participate in the community and contribute to quantum technologies as we have done in many traditional technologies."

Summary

"Materials science has a central role in the emerging quantum information technology industry."

"Both applied and fundamental opportunities in the field provide a range of contributions that materials scientists can make in a variety of materials systems, including superconductors, semiconductors, insulators in thin films, and bulk."

"Surfaces and interfaces are important, particularly between dissimilar materials."

"The largest contribution that materials researchers can make is to enhance our understanding of materials at cryogenic temperatures with energy resolution on the μeV scale."

Quantum Information Processing in the Neighborhood of a Black Hole [248]

This is a machine-generated summary of:

Crowder, Tanner; Lanzagorta, Marco: Quantum information processing in the neighborhood of a black hole [248].

Published in: Natural Computing (2019).

Link to original: https://doi.org/10.1007/s11047-019-09737-7

If you want to cite the papers, please refer to the original.

For technical reasons we could not place the page where the original quote is coming from.

Abstract-Summary

"Using an imperfectly prepared state, we show that in relativistic settings, the evolution of a massive spin-1/2 particle violates many standard assumptions made in quantum information theory, including complete positivity."

"Unlike other recent endeavors in relativistic quantum information, we are able to quantify and maximize how much information can be transferred through such a quantum process by calculating its scope."

"We show that, surprisingly, relativistic noise can increase the amount of information that can be transferred, and in fact, even if the initial state is arbitrarily close to the completely mixed state, information can still be transferred perfectly."

Introduction

"To Peres et al. [249] and Peres and Terno [250], the local dynamics of our model are linear, which allows us to calculate the maximal amount of classical information that can be transferred with a quantum state undergoing a Lorentz transformation."

"When one sends classical information through a quantum channel, one chooses a basis of the state space to represent the classical bits."

"Martin's procedure for calculating scope explicitly uses the fact that the channel is non-expansive in the Bloch representation, and he assumes that the classical information is encoded in to pure states."

"We believe the effects we study here serve as an exploration into non-standard dynamics that occur when we discard some of the usual assumptions made in quantum information science."

"A quantum computer on Earth's surface is moving through a gravitational field, which causes the computational states to rotate slightly during each step of the process."

Wigner Angles in Relativity

"We should discuss the meaning of a quantum state of definite momentum and spin projection in curved spacetime."

"We need to define local inertial frames at each point of spacetime."

"The spin projection and associated quantization axes of a quantum state in curved spacetime are described by the inertial observers at each point of the tetrad field."

"The description of the spin state and associated quantization axes can only be given with respect to the local inertial frames in each point of spacetime as defined by the tetrad field."

"And for the sake of simplicity, we rewrite the quantum states asFurthermore, the state given bydenotes a superposition of two states of definite momentum p and q as observed in a local inertial frame defined by some tetrad field."

The Reduced Density Matrix in Relativity

"There have been discussions about the reduced density matrix and the appropriate measurement observables related to relativistic spin-1/2 particles (Céleri et al. [251]; Saldanha and Vedral [252]; Taillebois and Avelar [253]; Choi [254]; Bauke [255])."

"In Saldanha and Vedral [252], they argue that since after a Lorentz transformation the measurement statistics of a massive spin-1/2 particle can depend on the momentum, the spin and momentum are not independent variables, and therefore the reduced density matrix is meaningless."

"It is our contention that the results in Saldanha and Vedral [252] are more of a commentary on the observables associated with a relativistic Stern–Gerlach device, rather than the usefulness of the reduced density matrix."

"As long as one computes the reduced density matrix in the measurement device's frame of reference by properly transforming the initial state with a Wigner rotation, and as long as one uses the correct observables, there is no ambiguity about the measurement statistics predicted by the reduced density matrix (Céleri et al. [251]; Taillebois and Avelar [253]; Choi [254])."

Quantum Information in Relativity

"When x is not one of the two vectors from case (1), from the observer's perspective, f has decreased the entropy of the spin state almost everywhere."

"Though, when classical information is sent through a quantum process, states are prepared and measured in the same basis, and they are not measured with an idealized basis."

"To represent classical information with quantum states, one usually chooses a basis of the state space to represent a 0 and 1."

"The scope of a quantum channel is the range of classical capacities as one varies over every basis in the state space (Martin [256])."

"In Martin [256], Martin specifically uses the fact that the map is non-expansive; however, non-expansivity is only used to show that the quantum channel induces a binary symmetric classical channel."

"By showing that the relativistic channels map our state space to the unit sphere, the quantum channel still induces a binary symmetric classical channel."

Kinematic and Gravitational Noise

"The Wigner angle never produces enough of an effect to greatly improve the capacity and, in this case, is more likely to be detrimental to information processing."

"The Wigner angle is completely dependent on the gravitational field generated by that black hole and the trajectory of the particle (Lanzagorta [257]; Alsing et al. [258]; Terashima and Ueda [259])."

"To obtain the entire range of values for the Wigner angle, we would need to explore the behavior of quantum information in the presence of strong gravitational fields, i.e., in the neighborhood of a black hole."

"The states would certainly always be more pure than the initial state; however, it's conceivable that the large contribution of the Wigner angle could be detrimental to a quantum process, since an increase in purity does not necessarily correspond to an increase in informatic content."

Conclusion

"(off that set of measure zero) the simple act of preparing and measuring the spin of such a state on Earth or on an orbiting satellite is not a completely positive process; albeit, this effect would be small and difficult to measure."

"By only measuring the spin state, the information about how the spin and momentum are mixed is destroyed."

"In a truncated version of this paper presented at Unconventional Computation and Natural Computation 2018 (Crowder and Lanzagorta [260]), we opined about whether one could employ a steganographic procedure to transmit information to someone in a different reference frame."

"Since in the preparer's frame of reference, the state is close to the completely mixed state, it would look like random noise to anyone trying to intercept the message in that frame, regardless of the measurement basis used."

Modern Physics of the Condensed State: Strong Correlations and Quantum Topology [261]

This is a machine-generated summary of:

Irkhin, V. Yu.; Skryabin, Yu. N.: Modern Physics of the Condensed State: Strong Correlations and Quantum Topology [261].

Published in: Physics of Metals and Metallography (2019).

Link to original: https://doi.org/10.1134/s0031918x19060061

If you want to cite the papers, please refer to the original.
For technical reasons we could not place the page where the original quote is coming from.

Abstract-Summary
"The theme of this survey is the application of new ideas of uncommon quantum states to the physics of the condensed state, in particular, of solids, in the context of the contemporary field theory."

"The variational approaches and the concepts of auxiliary particles, the corresponding mean-field approximations, the theory of gauge fields, the problem of confinement–deconfinement, the breakdown of the Fermi-liquid behavior, and exotic non-Fermi liquid states are considered."

"A survey of the contemporary theory of the entangled topological states, formation of spin liquid, strings, and string networks is given."

Introduction

"In the time that passed since the pioneer works of D. Haldane, M. Kosterlitz, and D. Thouless, considerable progress was achieved in the physics of the condensed state connected with the application of new, substantially quantum topological concepts, such as topological phases of substance, including those protected by symmetry, spin liquids, exotic excitations, strings, tensor networks, etc. These achievements were connected with the activities of several groups of researchers: Anderson [262, 263], Wen [264, 265], Sachdev [266], Coleman [267]."

"The low-energy effective theory of the topologically ordered state is called the topological quantum field theory (TQFT [268])."

"From a microscopic viewpoint the topological order is determined by the state of quantum spin liquid (QSL) with gapped energy spectrum [269, 270], which cannot be represented by the product of single-particle eigenstates if we do not take into account phase transitions with a closing of the energy gap."

Development of the Ideas and Methods of the Many-Electron Theory

"It makes it possible to rather simply describe the non-quasi-particle states caused by electron–magnon scattering, which previously were considered in the many-electron representation of Hubbard operators (see [271]) (in this case, the p^- operators describe the magnons, and f, describe the spinons."

"According to Anderson [272], the t–J Hamiltonian is not only a convenient alternative of the Hubbard model; it reflects the physical fact that the low-energy states are located in the subspace, which is overfilled and is described by one band of electronic states, since the anti-bonding states of doubles from the upper part of the band are neglected."

"The described picture is close to the hybridization two-band model of Kondo lattices, where the localized and itinerant states are initially separated, but are mixed up in the mean-field approximation for the auxiliary f pseudo-fermions, so that the latter participate in the formation of the Fermi surface [273]."

Fermi Liquid and Non-FERMI Liquid Phases

"A number of ideas were proposed on the appearance of uncommon metallic phases with a non-Fermi liquid behavior; in particular, such a metal can represent a fractionalized state with spinon and holon excitations."

"In the confinement phase, the entire picture of spin liquid is destroyed upon the consideration of fluctuation effects as a result of the interaction of a spinon with a gauge boson because of the proliferation of topological defects described by the corresponding homotopic groups [274]."

"Since the volume of the Fermi surface is retained, it cannot smoothly evolve from small to large state; therefore, the spin-liquid metal and the heavy Fermi liquid must be separated by a quantum phase transition at the zero temperature."

"In the antiferromagnetic and spin-liquid (disordered) phases the spinons are described as Schwinger bosons, and in the Fermi-liquid phase, as fermions (see [44])."

Theory of Quantum Phase Transitions. Spin Liquid and the Topological Order

"There is an unstable fixed point $s = s_c$ located on the line $\lambda_4 = 0$, separating the lines of renormalization-group trajectories in the direction of the magnetically ordered Néel state from the lines in the direction of paramagnetic spin liquid with a gapless U(1) photon and gapped spinon excitations."

"Such deconfinement phases are rather labile states of matter and can be unstable with respect to confinement phases with the usual order; for example, the instability of the U(1) spin liquid to the transition into the VBS."

"In the conducting magnets, upon a change in the parameters of interaction, additional quantum phase transitions appear in comparison with the Heisenberg model—the appearance of magnetic ordering in the ground state, a qualitative change in the behavior of the magnetic moment, etc. In the limit of strong correlations, such transitions are not described by simple approaches of the Stoner or Hertz–Millis type [275, 276]."

Lattice Gauge Theories and Strings

"The physical space of states of the Abel lattice gauge theory consists of closed loops of the electric flux."

"The physical space of states is locally gauge-invariant and satisfies the Gauss law Therefore, only closed loops of the electric flux are permitted: the theory does not have sources or sinks."

"The Abelian lattice gauge theory can be considered similarly—the physical space of states of the theory consists of the closed loops of the electric flux."

"Let us enumerate again the initial models, in which there are weak gauge interactions, deconfinement is possible, and a gauge boson and fermions appear: (1) The lattice Z_2 theory, which in the case of solids is dual to the Ising model having classical phase transitions."

"In the theory of string nets the intranet gauge bosons and fermions occur from qubits that form the space, and the "string net" is simply a model of qubits, the name, which indicates how the qubits are organized in the ground state."

Conclusions

"The topological phases, just as any other quantum systems, are characterized by an enormous space of states, and the addition of only one element can substantially change the state of the entire system (see, e.g. the Anderson's catastrophe of orthogonality [263, 277])."

"The topological order and entanglement lead to a number of new states of a quantum matter and to new physical phenomena, such as fractional charge, fractional and non-Abelian statistics, etc. If we could realize a quantum liquid consisting of oriented strings in real materials, this would allow us to create artificial elementary particles, an artificial world in artificial vacuum [264, 278]."

"Besides the physics of elementary particles, at present there are widely used formal methods, which unite the theory of topological states with the theory of gravity and structure of the universe, including holographic models (AdS/CFT theory) [279, 280]."

[Section 2]

"Quantum Phases and the Concept of Quantum Topology."

"DEVELOPMENT OF THE CONCEPTS AND METHODS OF MANY-ELECTRON THEORY."

"Quantum Phase Transitions and Incoherent States in Conducting Magnets."

Quantum Programming Language: A Systematic Review of Research Topic and Top Cited Languages [281]

This is a machine-generated summary of:

Garhwal, Sunita; Ghorani, Maryam; Ahmad, Amir: Quantum Programming Language: A Systematic Review of Research Topic and Top Cited Languages [281].

Published in: Archives of Computational Methods in Engineering (2019).

Link to original: https://doi.org/10.1007/s11831-019-09372-6

If you want to cite the papers, please refer to the original.

For technical reasons we could not place the page where the original quote is coming from.

Abstract-Summary

"Quantum programming is an emerging area developed in last 2 decades from the multidisciplinary research on quantum computing."

"This survey paper briefly gives an overview of state of the art in the field of quantum programming languages."

"This paper focuses on actual high-level quantum programming languages for quantum computers, their features and comparisons."

Introduction

"Quantum computing is generally studied at the hardware level, and its algorithms are usually represented using quantum Turing machines, quantum gates and circuits."

"Research in quantum computing accelerated after the development of Shor's algorithm for factorization of a large number [282]."

"A number of tools/simulators have been developed in the well-known classical languages such as C, C++, Java, MATLAB, Maple and Mathematica for implementing quantum algorithms and circuits."

"Developed algorithms are examined using the quantum circuit or partly pseudo code."

"To examine the efficiency and correctness of the algorithm, we need to execute the developed algorithms using quantum programming languages."

"There is a need to carry out a survey which includes the recent researches and compare it the development before 2008 in the area of quantum programming languages."

"Every quantum language must be simple, powerful, technology independent, and it can be executed on classical computers using simulator [283]."

Review Criterion

"RQ1: What are the different types of Quantum Programming Languages?"

"RQ2: What are the recent trends in the development of quantum programming languages?"

"RQ3: Which major Companies, Groups, Institutes and Universities are working in the development of new quantum languages?"

"RQ5: What are the most cited papers in the area of a quantum programming language?"

"RQ1 is a fundamental question for obtaining an overview of state of the art of quantum programming languages."

"RQ2 is aimed to obtain the recent developments in the area of quantum programming languages."

"RQ5 aimed to refer few most cited papers in the area of quantum programming language."

"Authors agree that number of citations criterion will be favourable for the quantum programming language which was developed earlier, but still, it will be useful in finding the most impactful research in this area."

Quantum Imperative Programming Language

"QCL is a full-fledged language of classical operations (control structure, elementary and structured data types) augmented on quantum types [284]."

"QCL provides quantum data types (qubit registers) and their functions for manipulations."

"LanQ [285, 286] is a high-level quantum programming language developed by Mlnarik at Masaryk University, Brno in 2006."

"It includes a compiler for quantum while language (Pure quantum language without classical variables) and having tools for optimizing quantum circuit, simulating quantum computation, analysis and verification of quantum programs."

"A program written in Q|SI > is compiled with f-Quantum Assembly Language with feedback (f-QASM)."

"Quantum assembler (qASM) [287] can be used for creating D-wave programs."

"Inversion, reordering of qubits, control operator and composition operators are available in Q. It is an object-oriented language used for quantum computing."

"In a narrow sense, Q is not a quantum programming language, but its library is written in C++."

Quantum Functional Language

"Abstract syntax [288] for the set of types for quantum lambda calculus is as follow: Here, operator!"

"Programming language specifications expressed using Quantum Lambda Calculus can be compared."

"Quantum Flow Chart (QFC) and Quantum Programming Language (QPL) are two quantum functional languages developed by Selinger [289] at the University of Ottawa in 2003."

"In QFC, quantum programs are represented using flowchart syntax with the functional flavour, whereas in QPL syntactic structure of quantum programs are represented using textual representation."

"It is a scalable, expressive and higher-order functional quantum programming language."

"Quipper consists of a larger class of circuit for qubit initialization, qubit termination, measurements of classical and quantum gates."

"It allows programmers to create classical functions on quantum data after making input and output data quantizable (data represented in the memory of classical computers as well as qubits within logical quantum circuits)."

Quantum Circuit Language

"Quantum circuit languages are developed for the design of quantum circuits."

"QuECT allows embedding circuit design in classical host language such as Java."

"Chakraborty [290] enhanced Java host language by enabling to handle quantum circuit design."

"QWIRE is a small quantum programming language for defining quantum circuits and embedded into Coq theorem prover."

"In QWIRE, quantum circuit language works independently from the host language concept."

"It is a highly modular and minimal circuit language."

"It provides a platform for high-level quantum computing with the expressiveness of embedded languages like Quipper and LIQUi| > ."

"The Dynamic feature of QWIRE makes it possible that the measurement result from the quantum circuit is processed dynamically as classical data inside Coq."

"Change in host language does not force any change in circuit language."

"The concept of box operator and unbox operators are used with host language and another circuit, respectively."

Multi-Paradigm Languages

"Quantum multi-paradigm languages are developed to support various features related to machine learning, design and optimization of circuits for performing computation such as photonic computation."

"Strawberry Fields is an open source quantum programming architecture developed by Killoran and others [291] for photonic quantum computing."

"It consists of a full-stack library for design, simulation, optimization, and quantum machine learning."

"Its front end consists of Strawberry fields API and Blackbird quantum programming language (A quantum assembly language for basic continuous-variable states, gates and measurement)."

"Qubit and Qumode are defined by It is an open quantum system developed by Krämer and others [292] using Julia programming language at Institut für Theoretische Physik, Universität Innsbruck, Austria in 2018."

"Q# [293] is a domain-specific programming language developed for quantum computation."

"Q# was developed by Quantum Architecture and Computation group (QuArC) at Microsoft in 2017."

Quantum Object Oriented Programming Language

"It is an extension of Feather-weight Java by adding the feature for handling quantum data and operations using a monadic layer."

"It is a small calculus for providing a formal semantics for Java language features."

"FJQuantum language [294] has the following feature to handle quantum concepts: 1."

"It provides the features to represent classes, attributes, methods, inheritance and dynamic casts with similar semantics as Feather-weight Java."

"It provides the features to create quantum states (monadic mreturn operator)."

Recent Trends and Results

"It has been observed that most of the researchers working in the area of quantum programming language prefer to publish on arxiv.org."

"We had considered papers from Journals named as SIAM Journal of Computing, The Computer Journal, International Journal of Theoretical Physics, ACM Transactions on Programming Languages and Systems, Acta Informatica, IBM Journal of Research and Development, Scientia Sinica Informationis, Physical Review Letters, International Journal of Quantum Information, Mathematical Structures in Computer Science and Computer Physics Communications."

"Although there are many research groups working in the area of quantum programming languages."

"We can see more research and development in the area of quantum programming languages in next 15–20 years."

"Most of the quantum programming languages (such as QCL, QFC, QPL, QML, Lambda Calculus, Q) are based on the QRAM model of computation."

"Improvement can be made on an already designed compiler for quantum programming language."

Classical Leakage Resilience from Fault-Tolerant Quantum Computation [295]

This is a machine-generated summary of:

Lacerda, Felipe G.; Renes, Joseph M.; Renner, Renato: Classical Leakage Resilience from Fault-Tolerant Quantum Computation [295].

Published in: Journal of Cryptology (2019).

Link to original: https://doi.org/10.1007/s00145-019-09310-6

If you want to cite the papers, please refer to the original.

For technical reasons we could not place the page where the original quote is coming from.

Abstract-Summary

"We establish a connection between leakage resilience and fault-tolerant quantum computation."

"We first prove that for a general leakage model, there exists a corresponding noise model in which fault tolerance implies leakage resilience."

"We show how to use constructions for fault-tolerant quantum computation to implement classical circuits that are secure in specific leakage models."

Introduction

"We take a novel approach to leakage resilience: fault-tolerant quantum computation."

"Since a fault-tolerant quantum computation must protect against phase errors (as well as more conventional bit-flip errors), it is necessarily leakage-resilient."

"Achieving leakage resilience in this way would require a fault-tolerant quantum circuit."

"We give a method for constructing a leakage-resilient classical circuit by modifying an appropriate fault-tolerant quantum circuit."

"Using the construction, one can transform an arbitrary classical circuit into a circuit that is resilient to leakage arising in any model for which reliable quantum computation is possible under the corresponding phase noise model."

"We stress that rather than presenting a specific leakage-resilient scheme, the contribution of this work aims to provide a novel approach to leakage resilience and connect this field of cryptography to the research area of fault-tolerant quantum computation."

Definitions

"Of a resource, we can define a private channel between honest parties A and B subject to possible eavesdropping by E as a resource that takes inputs from A and outputs them at B. Since the resource gives E no outputs, the channel is private by definition."

"That this holds whether E acts honestly or dishonestly; in abstract cryptography, the broader goal is to emulate the behavior of ideal resources in all situations, not just when assuming certain parties are honest and others dishonest."

"In the security arguments, we will also consider converters applied to the ideal resource by parties acting dishonestly."

"In the example, this assumption is ensured by the way the ideal resource is defined, since the only way information about y could leak to Eve is through the ciphertext f(x, y)."

Leakage Models and Quantum Noise

"The following lemma relates classical leakage models and quantum noise models."

"We will prove that such a simulator exists by showing that for any step of the computation, the leakage received by E is essentially independent of the circuit's current state (that is, the intermediate values encoded by its wires)."

"Consider an arbitrary timestep in the circuit."

From Quantum to Classical Circuits

"Due to the fact that the components have a classical translation, we also have a leakage-resilient classical gate."

"In order to show the equivalence between the quantum and classical circuits, we analyze each component in the following scenario: We assume that the quantum component has Z basis inputs (i.e., the inputs are classical), and that after execution, the outputs are measured in the Z basis."

"We then show that for each component there exists a classical circuit that, when given the same inputs, gives the same outputs as the quantum component after measuring."

"Since the components have a classical translation, the gadgets and the Toffoli gate have one as well."

"This is achieved by having, for each component, a classical translation that has the same wires as the quantum component."

Leakage-Resilient Gadgets

"Our fault-tolerant construction follows [296], which works in the model of independent phase noise where each wire in the circuit is subject to a phase error with probability p. This phase noise model is related to the independent leakage model, where each wire in the circuit leaks with probability 2p."

"In order to construct a leakage-resilient compiler, we just need to implement a set of fault-tolerant gates that is universal for classical computation."

"As we will see, our fault-tolerant implementation of the Toffoli gate involves only X, Z and CNOT gates, state preparation, measurement in the X and Z bases, and error correction."

"We present the fault-tolerant gadgets we use in the construction of the Toffoli gate."

"The usual way to prepare an ancilla state fault-tolerantly is to perform a verification step after encoding, where the state is rejected and the procedure is repeated if the verification detects too many errors."

Discussion

"We analyzed the independent leakage model, which corresponds to the independent phase noise model."

"A possible further direction is to take a leakage model that is used in other leakage resilience proposals and try to understand the corresponding noise model."

"We developed a concrete implementation of universal leakage-resilient computation based on the fault-tolerant construction of [296]."

"Taking into account the fact that we only want leakage resilience rather than fault tolerance, this threshold can be improved."

"While we do not expect existing results on fault tolerance to give direct constructions for the leakage models commonly studied in the literature (especially since fault tolerance has only been shown to work in very few noise models), we note that quantum fault tolerance is a stronger requirement than classical leakage resilience: As we have seen in this work, translating from the former to the latter allows us to make several simplifications."

Stationary States of a Dissipative Two-Qubit Quantum Channel and their Applications for Quantum Machine Learning [297]

This is a machine-generated summary of:

Ghasemian, E.: Stationary states of a dissipative two-qubit quantum channel and their applications for quantum machine learning [297].

Published in: Quantum Machine Intelligence (2023).

Link to original: https://doi.org/10.1007/s42484-023-00096-2

If you want to cite the papers, please refer to the original.

For technical reasons we could not place the page where the original quote is coming from.

Abstract-Summary

"We address how to control and harness these unwanted effects that arise from the coupling of a system with its environment, to provide stationary entangled states for quantum machine learning (QML)."

"To do so, we design a dissipative quantum channel, i.e., a two-qubit system interacting with a squeezed vacuum field reservoir, and study the output state of the channel by solving the corresponding master equation, especially, in the small squeezing regime."

"We show that the time-dependent output state of the channel is the so-called two-qubit X-states that generalize many families of entangled two-qubit states."

"By considering a general Bell diagonal state as the initial state of the system, we reveal that this dissipative channel generates two well-known classes of entangled mixed state and Werner-like states in the steady-state regime."

"This channel provides an efficient way to determine whether a given initial state results in a stationary entangled state or not."

"In this line, we propose a general theoretical scheme for quantum neural networks (QNNs) implemented with the variational quantum circuits, which encode data in continuous variables (CVs) of the two-qubit states."

"The linear (also referred to as affine) and nonlinear (activation function) transformations are enacted in the QNN using the stationary states of the two-qubit channel and measurement process, respectively."

Introduction

"Dissipation can be exploited as a fully fledged quantum resource for universal quantum computation without any coherent dynamics needed to complement it."

"The so-called dissipative quantum computing is of fundamental interest for QNN research, since it allows quantum computing algorithms based on dynamic attractors and steady states (Schuld and others [298]; Rabinovich and others [299])."

"The incompatibility between the nonlinear, dissipative dynamics of the NNs and the linear, unitary dynamics of coherent quantum computing may be addressed through dissipative quantum computing."

"Dissipative quantum computing is established based on the theory of open quantum systems."

"The idea of open QNNs has been put forward based on dissipative quantum computing (Schuld and others [298])."

"In analogy with the universal quantum processor that can produce any desired two-qubit unitary transformation, our dissipative quantum channel may be implemented for dissipative quantum computation based on the non-unitary transformations."

Dissipative Dynamics of a Two-Qubit System

"We study the dynamical evolution of a two-qubit system under the influence of a squeezed vacuum field reservoir."

"We solve the corresponding quantum master equation and find its solution, i.e., the density matrix of the two-qubit system."

"In the continuation, we design a dissipative quantum channel using a two-qubit system and a squeezed vacuum field reservoir."

"We demonstrate when the two qubits interact globally with the reservoir, the system is driven to its stationary state."

"Speaking, it can be shown that a squeezed vacuum reservoir considerably enhances the non-classical properties of a quantum system with respect to a thermal reservoir (Ji and Liu [300]; Kowalewska-Kudłaszyk and Leoński [301]; Ghasemian and Tavassoly [302])."

"We propose a new practical approach for the generation of this well-known class of entangled state via the interaction of a two-qubit system with a squeezed vacuum field reservoir."

Steady-State Density Matrix

"We intend to find the stationary state of the considered dissipative two-qubit channel."

"We can state that the channel cannot protect the initial coherence and entanglement for these initial states, so they are classified into the first class of initial states."

"We showed that the entanglement of the output state of the system can be obtained by evaluating the parameter Q which depends on the initial state of the two-qubit system."

"This dissipative two-qubit channel provides an efficient way to determine whether a given initial state results in a stationary entangled state or not."

"This is an outstanding feature of a dissipative two-qubit channel since the stationary entangled states can be generated just by tuning the free parameters of a generic initial state."

"The dissipative channel provides the stationary entangled states with the maximum degree of entanglement and facilitates the required quantum resources for realistic quantum information processing and dissipative quantum computation."

Quantum Coherence

"Since the coupling to the environment drives the system to a steady state, thus, the state $\rho(t)$ asymptotically approaches some steady states ρ_{ss} which can be obtained by the solution of the following stationary equation (Khrennikov [303]) Our analysis shows that the stationary solution of the system is not unique, especially, in the small squeezing regime, i.e., it depends on the initial conditions."

"In the steady-state regime, the quantum coherence reduces as which depends on the initial state of the two-qubit system."

"It can be seen that the time evolution of quantum coherent strongly depends on both parameters, i.e., the squeezing parameter and initial state."

"For zero squeezing ($r = 0$), the system gradually loses its initial coherence in the first three plots, but in the last plot quantum coherence gradually increases and finally the system reaches its maximum coherence in the steady-state regime."

Machine Learning Based on Dissipative Two-Qubit Channel

"We describe how a quantum classifier can be built based on the stationary state of the two-qubit channel."

"We examine the potential application of a dissipative quantum channel for supervised ML by detecting the stationary entangled states of the quantum channel."

"We follow the latter approach and design a variational quantum circuit for CV-QML by encoding data into the initial states of a two-qubit system."

"As an interesting application, we address how the considered dissipative two-qubit channel can be exploited as a quantum classifier (QNN)."

"As can be found, the classifier based on the dissipative two-qubit channel can solve these problems with relatively high accuracies as the entanglement classification."

"A dissipative quantum channel is a promising candidate for simulating a CNN because it provides stationary states via its highly dependent dynamics on the initial conditions (Schuld and others [298])."

Summary and Conclusion

"We have shown that these unwanted effects that arise through the coupling of a system with its environment can be exploited for the generation of stationary entangled states and the realization of meaningful dissipative quantum computation."

"By considering a generic initial state for the system, we have shown that the channel generates the so-called X-type states that generalize a number of entangled states such as the Bell state and Werner-like state."

"Measuring these entangled states, i.e., Bell states are critical for performing many of the quantum communication protocols and entanglement distribution across a quantum network (Duan and Kimble [304]; Zhao and others [305])."

"The channel can be exploited to generate stationary entangled states."

"Entanglement detection can be established provided that the output state of the channel is measured."

"To do so, we have designed a quantum circuit based on some dissipative channels and electro-optical components."

References

1. Bukbech S, El Anouz K, El Allali Z, Metwally N, El Allati A (2023) Multiparameter estimation for a two-qubit system coupled to independent reservoirs using quantum Fisher information. Quant Stud: Math Found. https://doi.org/10.1007/s40509-023-00303-6
2. Genoni MG, Paris MGA, Adesso G, Nha H, Knight PL, Kim MS (2013) Optimal estimation of joint parameters in phase space. Phys Rev A 87:012107
3. Monras A, Illuminati F (2011) Measurement of damping and temperature: precision bounds in Gaussian dissipative channels. Phys Rev A 83:012315
4. Szczykulska M, Baumgratz T, Datta A (2016) Multi-parameter quantum metrology. Adv Phys X 1:621
5. Braunstein SL, Caves CM (1994) Statistical distance and the geometry of quantum states. Phys Rev Lett 72:3439
6. Metwally N (2011) Quantum dense coding and dynamics of information over Bloch channels. J Phys A: Math Theor 44:055305
7. Metwally N (2013) Teleportation of accelerated information. J Opt Soc Am B 30:233
8. Paris MG (2009) Quantum estimation for quantum technology. Int J Quant Inf 7:125
9. Safránek D (2017) Discontinuities of the quantum Fisher information and the Bures metric. Phys Rev A 95:052320
10. Slater, P.B.: Quantum Fisher–Bures information of two-level systems and a three-level extension. J Phys A 29, 271 (1996).
11. Sommers HJ, Zyczkowski K (2003) Bures volume of the set of mixed quantum states. J Phys A 36:10083
12. Wootters WK (1981) Statistical distance and Hilbert space. Phys Rev D 23:357

13. Polychronopoulou A, Alshehri J, Obradovic Z (2023) Quantum-inspired measures of network distinguishability. Soc Netw Analy Mining. https://doi.org/10.1007/s13278-023-01069-w
14. Bunke H, Dickinson PJ, Kraetzl M, Wallis WD (2007) A graph theoretic approach to enterprise network dynamics. Birkhäuser, Boston
15. Biamonte J, Faccin M, De Domenico M (2019) Complex networks from classical to quantum. Commun Phys 2(1):1–10
16. Javarone MA, Armano G (2013) Quantum-classical transitions in complex networks. J Stat Mech Theory Exp 2013(04): P04019.
17. Polychronopoulou A, Alshehri J, Obradovic Z (2021) Distinguishability of graphs: a case for quantum-inspired measures. In: Proceedings of the 2021 IEEE/ACM International conference on advances in social networks analysis and mining.
18. Sánchez-García RJ, Cozzo E, Moreno Y (2014) Dimensionality reduction and spectral properties of multilayer networks. Phys Rev E 89(5):052815
19. Song D, Bauman NP, Prawiroatmodjo G, Peng B, Granade C, Rosso KM, Low GH, Roetteler M, Kowalski K, Bylaska EJ (2023) Periodic plane-wave electronic structure calculations on quantum computers. Mater Theory. https://doi.org/10.1186/s41313-022-00049-5
20. Szabo A, Ostlund NS (2012) Modern quantum chemistry: introduction to advanced electronic structure theory. Courier Corporation, North Chelmsford
21. Ross IG (1952) Calculations of the energy levels of acetylene by the method of antisymmetric molecular orbitals, including σ - π interaction. Trans Faraday Soc 48:973–991. https://doi.org/10.1039/TF9524800973
22. Gan Z, Grant DJ, Harrison RJ, Dixon DA (2006) The lowest energy states of the group-IIIA–group-va heteronuclear diatomics: BN, BP, AlN, and AlP from full configuration interaction calculations. J Chem Phys 125(12):124311. https://doi.org/10.1063/1.2335446
23. McArdle S, Endo S, Aspuru-Guzik A, Benjamin SC, Yuan X (2020) Quantum computational chemistry. Rev Mod Phys 92(1):015003. https://doi.org/10.1103/RevModPhys.92.015003
24. Tubman NM, Freeman CD, Levine DS, Hait D, Head-Gordon M, Whaley KB (2020) Modern approaches to exact diagonalization and selected configuration interaction with the adaptive sampling CI method. J Chem Theory Comput 16(4):2139–2159. https://doi.org/10.1021/acs.jctc.8b00536
25. Sugisaki K, Sakai C, Toyota K, Sato K, Shiomi D, Takui T (2021) Quantum algorithm for full configuration interaction calculations without controlled time evolutions. J Phys Chem Lett 12(45):11085–11089. https://doi.org/10.1021/acs.jpclett.1c03214
26. Kawashima Y, Lloyd E, Coons MP, Nam Y, Matsuura S, Garza AJ, Johri S, Huntington L, Senicourt V, Maksymov AO et al (2021) Optimizing electronic structure simulations on a trapped-ion quantum computer using problem decomposition. Commun Phys 4(1):1–9. https://doi.org/10.1038/s42005-021-00751-9
27. Coester F (1958) Bound states of a many-particle system. Nucl Phys 7:421–424. https://doi.org/10.1016/0029-5582(58)90280-3
28. Coester F, Kummel H (1960) Short-range correlations in nuclear wave functions. Nucl Phys 17:477–485. https://doi.org/10.1016/0029-5582(60)90140-1
29. Čížek J (1966) On the correlation problem in atomic and molecular systems. Calculation of wavefunction components in ursell-type expansion using quantum-field theoretical methods. J Chem Phys 45(11):4256–4266. https://doi.org/10.1063/1.1727484
30. Paldus J, Čížek J, Shavitt I (1972) Correlation problems in atomic and molecular systems. IV. Extended coupled-pair manyelectron theory and its application to the BH_3 molecule. Phys Rev A 5:50–67. https://doi.org/10.1103/PhysRevA.5.50
31. Mukherjee D, Moitra RK, Mukhopadhyay A (1975) Correlation problem in open-shell atoms and molecules: a non-perturbative linked cluster formulation. Mol Phys 30(6):1861–1888. https://doi.org/10.1080/00268977500103351
32. Purvis GD, Bartlett RJ (1982) A full coupled-cluster singles and doubles model: the inclusion of disconnected triples. J Chem Phys 76(4):1910–1918. https://doi.org/10.1063/1.443164

33. Bishop RF, Kümmel H (1987) The coupled-cluster method. Phys Today 40(3):52. https://doi.org/10.1063/1.881103
34. Paldus J, Li X (1999) A critical assessment of coupled cluster method in quantum chemistry. Adv Chem Phys 110:1–175. https://doi.org/10.1002/9780470141694.ch1
35. Crawford TD, Schaefer HF (2000) An introduction to coupled cluster theory for computational chemists. Rev Comput Chem 14:33–136. https://doi.org/10.1002/9780470125915.ch2
36. Bartlett RJ, Musiał M (2007) Coupled-cluster theory in quantum chemistry. Rev Mod Phys 79:291–352. https://doi.org/10.1103/RevModPhys.79.291
37. Peng B, Bauman NP, Gulania S, Kowalski K (2021) Annu. Rep. Comput. Chem. Coupled cluster green's function: past, present, and future, vol 17. Elsevier, Amsterdam, pp 23–53. https://doi.org/10.1016/bs.arcc.2021.08.002
38. Green G (1854) An essay on the application of mathematical analysis to the theories of electricity and magnetism. J für die Reine und Angew Math (Crelles J) 1854(47):161–221. https://doi.org/10.1515/crll.1854.47.161
39. Green G (2014) Cambridge Library Collection. An essay on the application of mathematical analysis to the theories of electricity and magnetism (Mathematics). Cambridge University Press, Cambridge, pp 1–82
40. Feynman R (1949) The theory of positrons. Phys Rev 76:749–759. https://doi.org/10.1103/PhysRev.76.749
41. Feynman R (1948) Space-time approach to non-relativistic quantum mechanics. Rev Mod Phys 20:367–387. https://doi.org/10.1103/RevModPhys.20.367
42. Martin P, Schwinger J (1959) Theory of many-particle systems. I. Phys Rev 115:1342–1373. https://doi.org/10.1103/PhysRev.115.1342
43. Baym G, Kadanoff L (1961) Conservation laws and correlation functions. Phys Rev 124:287–299. https://doi.org/10.1103/PhysRev.124.287
44. Bylaska EJ, Song D, Bauman NP, Kowalski K, Claudino D, Humble TS (2021) Quantum solvers for plane-wave hamiltonians: abridging virtual spaces through the optimization of pairwise correlations. Front Chem 9:603019. https://doi.org/10.3389/fchem.2021.603019
45. Bylaska E, Tsemekhman K, Govind N, Valiev M (2011) Large-scale plane-wave-based density functional theory: formalism, parallelization, and applications. Comput Methods Large Syst Electron Struct Approaches Biotechnol Nanotechnol 77–116. https://doi.org/10.1002/9780470930779.ch3
46. Bylaska EJ, Tsemekhman K, Baden SB, Weare JH, Jonsson H (2011) Parallel implementation of γ-point pseudopotential plane-wave DFT with exact exchange. J Comput Chem 32(1):54–69. https://doi.org/10.1002/jcc.21598
47. Bylaska EJ (2017) Annual reports in computational chemistry. Plane-wave DFT methods for chemistry, vol 13. Elsevier, Amsterdam, pp 185–228. https://doi.org/10.1016/bs.arcc.2017.06.006
48. Bylaska EJ, Rosso K (2018) Corresponding orbitals derived from periodic bloch states for electron transfer calculations of transition metal oxides. J Chem Theory Comput 14(8):4416–4426. https://doi.org/10.1021/acs.jctc.7b01180
49. Bylaska EJ, Waters K, Hermes ED, Zádor J, Rosso KM (2020) A Filon-like integration strategy for calculating exact exchange in periodic boundary conditions: a plane-wave DFT implementation. Mater Theory 4(1):1–31. https://doi.org/10.1186/s41313-020-00019-9
50. Pickett WE (1989) Electronic structure of the high-temperature oxide superconductors. Rev Mod Phys 61(2):433. https://doi.org/10.1103/RevModPhys.61.433
51. Ihm J, Zunger A, Cohen ML (1979) Momentum-space formalism for the total energy of solids. J Phys C: Solid State Phys 12(21):4409. https://doi.org/10.1088/0022-3719/12/21/009
52. Car R, Parrinello M (1985) Unified approach for molecular dynamics and density-functional theory. Phys Rev Lett 55(22):2471. https://doi.org/10.1103/PhysRevLett.55.2471
53. Payne MC, Teter MP, Allan DC, Arias TA, Joannopoulos JD (1992) Iterative minimization techniques for ab initio total-energy calculations: molecular dynamics and conjugate gradients. Rev Mod Phys 64:1045–1097. https://doi.org/10.1103/RevModPhys.64.1045

54. Remler DK, Madden PA (1990) Molecular dynamics without effective potentials via the car-parrinello approach. Mol Phys 70(6):921–966. https://doi.org/10.1080/00268979000101451
55. Kresse G, Furthmüller J (1996) Efficient iterative schemes for ab initio total-energy calculations using a plane-wave basis set. Phys Rev B 54(16):11169. https://doi.org/10.1103/PhysRevB.54.11169
56. Marx D, Hutter J (2000) Modern methods and algorithms of quantum chemistry. Jülich, Germany, pp 301–449
57. Martin RM (2004) Electronic structure: basic theory and practical methods. (Cambridge University Press, Cambridge. https://doi.org/10.1017/CBO9780511805769
58. Valiev M, Bylaska EJ, Gramada A, Weare JH (2002) Reviews in modern quantum chemistry: a celebration of the contributions of Robert G. Parr. In: Sen KD (ed) First principles molecular dynamics simulations using density-functional theory. World Scientific, Singapore
59. Chen Y, Bylaska E, Weare J (2016) First principles estimation of geochemically important transition metal oxide properties. Mol Model Geochem React Introduction 107. https://doi.org/10.1002/9781118845226.ch4
60. King HF, Stanton RE, Kim H, Wyatt RE, Parr RG (1967) Corresponding orbitals and the nonorthogonality problem in molecular quantum mechanics. J Chem Phys 47(6):1936–1941. https://doi.org/10.1063/1.1712221
61. Adamowicz L, Bartlett RJ (1987) Optimized virtual orbital space for high-level correlated calculations. J Chem Phys 86(11):6314–6324. https://doi.org/10.1063/1.452468
62. Adamowicz L, Bartlett RJ, Sadlej AJ (1988) Optimized virtual orbital space for high-level correlated calculations. ii. electric properties. J Chem Phys 88(9):5749–5758. https://doi.org/10.1063/1.454721
63. Kendall RA, Aprà E, Bernholdt DE, Bylaska EJ, Dupuis M, Fann GI, Harrison RJ, Ju J, Nichols JA, Nieplocha J et al (2000) High performance computational chemistry: an overview of nwchem a distributed parallel application. Comput Phys Commun 128(1–2):260–283. https://doi.org/10.1016/S0010-4655(00)00065-5
64. Valiev M, Bylaska EJ, Govind N, Kowalski K, Straatsma TP, Van Dam HJ, Wang D, Nieplocha J, Aprà E, Windus TL et al (2010) NWChem: a comprehensive and scalable open-source solution for large scale molecular simulations. Comput Phys Commun 181(9):1477–1489. https://doi.org/10.1016/j.cpc.2010.04.018
65. Apra E, Bylaska EJ, De Jong WA, Govind N, Kowalski K, Straatsma TP, Valiev M, van Dam HJ, Alexeev Y, Anchell J et al (2020) NWChem: past, present, and future. J Chem Phys 152(18):184102. https://doi.org/10.1063/5.0004997
66. Shor PW (1995) Scheme for reducing decoherence in quantum computer memory. Phys Rev A 52:R2493–R2496. https://doi.org/10.1103/PhysRevA.52.R2493
67. Cory DG, Price MD, Maas W, Knill E, Laflamme R, Zurek WH, Havel TF, Somaroo SS (1998) Experimental quantum error correction. Phys Rev Lett 81:2152–2155. https://doi.org/10.1103/PhysRevLett.81.2152
68. Reed MD, DiCarlo L, Nigg SE, Sun L, Frunzio L, Girvin SM, Schoelkopf RJ (2012) Realization of three-qubit quantum error correction with superconducting circuits. Nature 482:382–385. https://doi.org/10.1038/nature10786
69. Bagherian M, Chehade S, Whitney B, Passian A (2023) Classical and quantum compression for edge computing: the ubiquitous data dimensionality reduction. Computing. https://doi.org/10.1007/s00607-023-01154-0
70. Nguyen TT, Ha VN, Le LB, Schober R (2019) Joint data compression and computation offloading in hierarchical fog-cloud systems. IEEE Trans Wirel Commun 19:293–309
71. Deepu CJ, Heng C-H, Lian Y (2016) A hybrid data compression scheme for power reduction in wireless sensors for IoT. IEEE Trans Biomed Circuits Syst 11(2):245–254
72. Passian A, Imam N (2019) Nanosystems, edge computing, and the next generation computing systems. Sensors 19(18):4048
73. Rozema LA, Mahler DH, Hayat A, Turner PS, Steinberg AM (2014) Quantum data compression of a qubit ensemble. Phys Rev Lett 113(16):160504

74. Renes JM, Renner R (2012) One-shot classical data compression with quantum side information and the distillation of common randomness or secret keys. IEEE Trans Inf Theory 58(3):1985–1991
75. Yan F, Venegas-Andraca SE, Hirota K (2022) Toward implementing efficient image processing algorithms on quantum computers. Soft Comput. https://doi.org/10.1007/s00500-021-06669-2
76. Yan F, Venegas-Andraca SE (2020) Quantum image processing. Springer, Berlin
77. Banaszek K, Cramer M, Gross D (2012–2013) Focus issue on quantum tomography (31 articles). New J Phys 14–15: 125020.
78. Youssry A, Ferrie C, Tomamichel M (2019) Efficient online quantum state estimation using a matrix-exponentiated gradient method. New J Phys 21:033006
79. Biamonte J, Wittek P, Pancotti N, Rebentrost P, Wiebe N, Lloyd S (2017) Quantum machine learning. Nature 549:195–202
80. Venegas-Andraca SE, Cruz-Santos W, McGeoch C, Lanzagorta M (2018) A cross-disciplinary introduction to quantum annealing-based algorithms. Contemp Phys 59(2):174–197
81. Cruz-Santos W, Venegas-Andraca SE, Lanzagorta M (2019) A QUBO formulation of minimum multicut problem instances in trees for D-Wave quantum annealers. Sci Rep 9:17216
82. Yu C, Gao F, Liu C, Huynh D, Reynolds M, Wang J (2019) Quantum algorithm for visual tracking. Phys Rev A 99:022301
83. Trugenberger CA (2002) Quantum pattern recognition. Quantum Inf Process 1(6):471–493
84. Yan F, Li N, Hirota K (2021) QHSL: a quantum hue, saturation, and lightness color model. Inform Sci 577:196–213
85. Li H, Fan P, Xia H, Peng H, Long G (2020) Efficient quantum arithmetic operation circuits for quantum image processing. Sci China Phys Mech Astron 63:280311
86. Wright J, Gowrishankar M, Claudino D, Lotshaw PC, Nguyen T, McCaskey AJ, Humble TS (2022) Numerical simulations of noisy quantum circuits for computational chemistry. Mater Theory. https://doi.org/10.1186/s41313-022-00047-7
87. Claudino D, Wright J, McCaskey AJ, Humble TS (2020) Benchmarking adaptive variational quantum eigensolvers. Front Chem 8:1152
88. Powell MJD (1998) Direct search algorithms for optimization calculations. Acta Numerica 7:287–336. https://doi.org/10.1017/S0962492900002841
89. Nocedal J (1980) Updating quasi-newton matrices with limited storage. Math Comp 35:773–782
90. Liu DC, Nocedal J (1989) On the limited memory BFGS method for large scale optimization. Math Program 45:503–528
91. Temme K, Bravyi S, Gambetta JM (2017) Error mitigation for short-depth quantum circuits. Phys Rev Lett 119:180509. https://doi.org/10.1103/PhysRevLett.119.180509
92. Ville J-L, Morvan A, Hashim A, Naik RK, Lu M, Mitchell B, Kreikebaum J-M, O'Brien KP, Wallman JJ, Hincks I et al (2021) Leveraging randomized compiling for the qite algorithm. arXiv preprint arXiv:2104.08785. https://doi.org/10.48550/ARXIV.2104.08785. https://arxiv.org/abs/2104.08785
93. de Jong E (2022) Own the unknown: an anticipatory approach to prepare society for the quantum age. Digi Soc. https://doi.org/10.1007/s44206-022-00020-4
94. WRR (2021) Opgave AI. De nieuwe systeemtechnologie. WRR, The Hague
95. Nielsen MA, Chuang IL (2011) Quantum computation and quantum information: 10th anniversary. Cambridge University Press, Cambridge
96. Krelina M (2021) Quantum technology for military applications. EPJ Quant Technol 8(1):24
97. Hoofnagle C, Garfinkel S (2022) Law and policy for the quantum age. Cambridge University Press
98. Der Derian J, Wendt A (2020) 'Quantizing international relations': the case for quantum approaches to international theory and security practice. Secur Dialogue 51(5):399–413

99. Combarro EF, Vallecorsa S, Rodríguez-Muñiz LJ, Aguilar-González Á, Ranilla J, Di Meglio A (2021) A report on teaching a series of online lectures on quantum computing from CERN. J Supercomput. https://doi.org/10.1007/s11227-021-03847-9
100. Guan W, Perdue G, Pesah A, Schuld M, Terashi K, Vallecorsa S, Vlimant JR (2020) Quantum machine learning in high energy physics. arXiv preprint arXiv:2005.08582
101. Sharma KK (2020) Quantum machine learning and its supremacy in high energy physics. Modern Phys Lett A:2030024
102. US Energy Information Administration (2021) In: Annual Energy Outlook 2019 with Projections to 2050. https://www.eia.gov/outlooks/aeo/ . Accessed 7 July 2021.
103. Fox MF, Zwickl BM, Lewandowski H (2020) Preparing for the quantum revolution: what is the role of higher education? Phys Rev Phys Educ Res 16(2)
104. Carrascal G, del Barrio AA, Botella G (2020) First experiences of teaching quantum computing. J Supercomput:1–30
105. LaRose R (2019) Teaching quantum computing through programming. Medium. https://medium.com/@rlarose_26759/teaching-quantum-computing-through-programming-799283c9769a
106. Mykhailova M, Svore KM (2020) Teaching quantum computing through a practical software-driven approach: experience report. In: Proceedings of the 51st ACM technical symposium on computer science education, pp 1019–1025
107. Salehi Ö, Seskir Z, Tepe İ (2020) Teaching quantum computing to an audience beyond physicists: a case study over 22 workshops in 10 countries. arXiv preprint arXiv:2010.13552
108. Tappert CC, Frank RI, Barabasi I, Leider AM, Evans D, Westfall L (2019) Experience teaching quantum computing. Assoc Support Comput Users Educ
109. Giani A, Eldredge Z (2021) Quantum computing opportunities in renewable energy. SN Computer Science. https://doi.org/10.1007/s42979-021-00786-3
110. Nigg SE, Lörch N, Tiwari R (2017) Robust quantum optimizer with full connectivity. Sci Adv 3:4
111. Bian T, Murphy D, Xia R et al (2019) Quantum computing methods for electronic states of the water molecule. Mol Phys 117(15–16):2069–2082
112. Kumar A, Garhwal S (2021) State-of-the-art survey of quantum cryptography. Archiv Computat Meth Eng. 10.1007/s11831-021-09561-2
113. Shor PW (1997) Polynomial-time algorithms for prime factorization and discrete logarithms on a quantum computer. SIAM J Comput 26:1484–1509
114. Gisin N, Ribordy G, Tittel W et al (2002) Quantum cryptography. Rev Mod Phys 74(1):145
115. Alleaume R, Branciard C, Bouda J, Debuisschert T, Dianati M, Gisin N, Godfrey M, Grangier P, Langer T, Lutkenhaus N, Monyk C, Painchault P, Peev M, Poppe A, Pornin T, Rarity J, Renner R, Ribordy G, Riguidel M, Salvail L, Shields A, Weinfurter H, Zeilinger A (2014) Using quantum key distribution for cryptographic purposes: a survey. Theor Comput Sci 560:62–81
116. Giampouris D (2016) Short review on quantum key distribution protocols. In: Vlamos P (ed) GeNeDis computational biology and bioinformatics, advances in experimental medicine and biology, vol 988. Springer, Cham, pp 149–157
117. Diamanti E, Lo HK, Qi B, Yuan Z (2016) Practical challenges in quantum key distribution. npj Quantum Inf 2:16025
118. Long GL (2017) Quantum secure direct communication: principles, current status, perspectives. In: 2017 IEEE 85th vehicular technology conference (VTC 2017 Spring) 4–7 June 2017 Sydney, Australia, pp. 1–5
119. Zhou T, Shen J, Li X, Wang C, Shen J (2018) Quantum cryptography for the future internet and the security analysis. Secur Commun Netw 8214619:1–7
120. Wootters WK, Zurek WH (1982) A single quantum cannot be cloned. Nature 299(5886):802–803. https://doi.org/10.1038/299802a0
121. Brandao FGSL, Oppenheim J (2012) The quantum one-time pad in the presence of an eavesdropper. Phys Rev Lett 108(4):040504

122. Griffet C (2019) From discrete-to continuous-variable protocols for quantum key distribution., Master Thesis,. Universite Libre De Bruxelles
123. Garcia-Patron R, Cerf NJ (2009) Continuous-variable quantum key distribution protocols over noisy channels. Phys Rev Lett 102:130501-1–130501-4
124. Cerf NJ, Grangier P (2007) From quantum cloning to quantum key distribution with continuous variables: a review (Invited). J Opt Soc Am 24(2):324–334
125. Leverrier A, Grangier P (2011) Continuous-variable quantum key distribution protocols with a discrete modulation. arXiv:1002.4083
126. Papanastasiou P, Pirandola S (2020) Continuous-variable quantum cryptography with discrete alphabets: composable security under collective gaussian attacks, pp. 1–6. arXiv:1912.11418
127. Qi B, Fung CHF, Lo HK, Ma X (2007) Time-shift attack in practical quantum cryptosystems. Quant Inf Comput 7(1):73–82
128. Huttner B, Imoto N, Gisin N, Mor T (1995) Quantum cryptography with coherent states. Phys Rev A 51(3):1863–1869
129. Lutkenhaus N (2000) Security against individual attacks for realistic quantum key distribution. Phys Rev A 61:052304–1–052304-10
130. Liu WT, Sun SH, Liang LM, Yuan JM (2011) Proof-of-principle experiment of a modifed photon-number-splitting attack against quantum key distribution. Phys Rev A 83:042326–1–042326-5
131. Fung CHF, Qi B, Tamaki K, Lo HK (2007) Phase-remapping attack in practical quantum-key-distribution systems. Phys Rev A 75(3):032314-1–032314-12
132. Xu F, Qi B, Lo HK (2010) Experimental demonstration of phase-remapping attack in a practical quantum key distribution system. New J Phys 12:113026
133. Sun SH, Xu F, Jiang MS, Ma XC, Lo HK, Liang LM (2015) Effect of source tampering in the security of quantum cryptography. Phys Rev A 92(2):022304
134. Lamas-Linares A, Kurtsiefer C (2007) Breaking a quantum key distribution system through a timing side channel. Opt Express 15(15):9388–9393
135. Makarov V, Hjelme DR (2005) Faked states attack on quantum cryptosystems. J Mod Opt 52:691–705
136. Zhao Y, Fung CHF, Qi B, Chen C, Lo HK (2008) Quantum hacking: experimental demonstration of time-shift attack against practical quantum-key-distribution systems. Phys Rev A 78:042333-1–042333-5
137. Wei K, Zhang W, Tang YL, You L, Xu F (2019) Implementation security of quantum key distribution due to polarization-dependent efficiency mismatch. Phys Rev A 100(2):022325
138. Garcia-Patron R, Wong FNC, Shapiro JH (2010) Optimal individual attack on BB84 quantum key distribution using single- photon two-qubit quantum logic Proc SPIE Int Soc Opt Eng 7702:77020C-1–77020C-10
139. Boyer B, Liss R, Mor T (2017) Security against collective attacks of a modified BB84 QKD protocol with information only in one basis. In: Proceedings of the 2nd international conference on complexity, future information systems and risk (COMPLEXIS 2017), vol 2, pp. 23–29
140. Chamoli A, Bhandari CM (2009) Secure direct communication based on ping-pong protocol. Quant Inf Process 8:347–356
141. Sarvaghad-Moghaddam M (2019) Efficient controlled bidirectional quantum secure direct communication using entanglement swapping in a network. arXiv:1902.11188 1–15
142. Tan YG, Cai QY (2008) Classical correlation in quantum dialogue. Int J Quant Inf 6(2):325–329
143. Krawec WO (2015) Mediated semi-quantum key distribution. Phys Rev A 91:032323
144. Yu KF, Yang CW, Liao CH, Hwang T (2014) Authenticated semi-quantum key distribution protocol using Bell states. Quant Inf Process 13:1457–1465
145. Zou X, Qiu D, Zhang S, Mateus P (2015) Semiquantum key distribution without invoking the classical party's measurement capability. Quant Inf Process 14:2981–2996

146. Tsai CW, Yang CW, Lee NY (2019) Semi-quantum secret sharing protocol using W-state. Mod Physi Lett A 34(27):1950213-1–1950213-12
147. Lin PH, Tsai CW, Hwang T (2019) Mediated semi-quantum key distribution using single photons. Ann Phys 531(8):1800347-1–1800347-7
148. Yan L, Sun YH, Chang Y, Zhang SB, Wan GG, Sheng ZW (2018) Semi-quantum protocol for deterministic secure quantum communication using Bell states. Quant Inf Process 17:315-1–315-12
149. Sun Z, Zhang C, Wang P, Yu J, Zhang Y, Long D (2016) Multi-party quantum key agreement by an entangled six-qubit state. Int J Theor Phys 55(3):1920–1929
150. Sun Z, Huang J, Wang P (2016) Efficient multiparty quantum key agreement protocol based on commutative encryption. Quant Inf Process 15:2101–2111
151. Huang W, Su Q, Liu B, He YH, Fan F, Xu BJ (2017) Efficient multiparty quantum key agreement with collective detection. Sci Rep 7:15264-1–15264-9
152. Cao WF, Zhen YZ, Zheng YL, Li L, Chen ZB, Liu NL, Chen K (2018) One-sided measurement-device-independent quantum key distribution. Phys Rev A 97:012313
153. He WT, Wang J, Zhang TT, Alzahrani F, Hobiny A, Alsaedi A, Hayat T, Deng FG (2019) High-efficiency three-party quantum key agreement protocol with quantum dense coding and Bell states. Int J Theor Phys 58:2834–2846
154. Tomamichel M, Fehr S, Kaniewski J, Wehner S (2013) One-sided Device-independent QKD and position-based cryptography from monogamy games, advances in cryptology-EUROCRYPT. In: 32nd annual international conference on the theory and applications of cryptographic techniques, Athens, Greece, May 26–30. Lecture notes in computer science (LNCS), vol 7881, pp. 609–625
155. Lo HK, Curty M, Qi B (2012) Measurement-device-independent quantum key distribution. Phys Rev Lett 108(13):130503
156. Tang Z, Wei K, Bedroya O, Qian L, Lo HK (2016) Experimental measurement-device-independent quantum key distribution with imperfect sources. Phys Rev A 93:042308
157. Pawlowski M, Brunner N (2011) Semi-device-independent security of one-way quantum key distribution. Phys Rev A 84(1):010302
158. Lim CCW, Korzh B, Martin A, Bussieres F, Thew R, Zbinden H (2014) Detector-device-independent quantum key distribution. Appl Phys Lett 105:221112
159. Gonzalez P, Rebon L, Silva TFD, Figueroa M, Saavedra C, Curty M, Lima G, Xavier GB, Nogueira WAT (2015) Quantum key distribution with untrusted detectors. Phys Rev A 92(2):022337
160. Sajeed S, Huang A, Sun S, Xu F, Makarov V, Curty M (2016) Insecurity of detector-device-independent quantum key distribution. Phys Rev Lett 117(25):250505
161. Bernstein DJ (2009) Introduction to post-quantum cryptography. In: Bernstein DJ, Buchmann J, Dahmen E (eds) Post-quantum cryptography. Springer, Berlin, pp 1–14
162. Hoffstein J, Pipher J, Silverman JH (1998) NTRU: a ring-based public key cryptosystem. In: International algorithmic number theory symposium ANTS 1998: algorithmic number theory, Lecture notes in computer science, vol 1423. LNCS, Springer, pp 267–288
163. Dods C, Smart NP, Stam M (2005) Hash based digital signature schemes. In: In: tenth proceeding of IMA international conference on cryptography and coding (IMACC 2005), Lecture notes in computer science, vol 3796. Springer, Berlin, pp 96–115
164. Hulsing A (2013) W-OTS$^+$-shorter signatures for hash-based signature schemes. In: Proceeding of 6th international conference on cryptology in Africa, Cairo, Egypt, June 22–24, Lecture notes in computer science. Springer, Berlin, Heidelberg, vol 7918, pp. 173–188
165. Amer O, Krawec WO (2019) Semiquantum key distribution with high quantum noise tolerance. Phys Rev A 100:022319-1–022319-16
166. Arrighi P, Salvail L (2006) Blind quantum computation. Int J Quant Inf 4(5):883–898
167. Broadbent A, Fitzsimons J, Kashefi E (2009) Universal blind quantum computation. In: 50*th* annual IEEE symposium on foundations of computer science. Atlanta, CA, USA 25–27 Oct, pp. 517–526

168. Qi B, Siopsis G (2015) Loss-tolerant position-based quantum cryptography. Phys Rev A 91:042337
169. Boaron A, Boso G, Rusca D, Vulliez C, Autebert C, Caloz M, Perrenoud M, Gras G, Bussieres F, Li MJ, Nolan D, Martin A, Zbinden H (2018) Secure quantum key distribution over 421 km of optical fiber. Phys Rev Lett 121:190502
170. Ma H, Govoni M, Galli G (2020) Quantum simulations of materials on near-term quantum computers. npj Comput Mater. https://doi.org/10.1038/s41524-020-00353-z
171. Georges A, Kotliar G, Krauth W, Rozenberg MJ (1996) Dynamical mean-field theory of strongly correlated fermion systems and the limit of infinite dimensions. Rev Mod Phys 68:13–125
172. Kotliar G et al (2006) Electronic structure calculations with dynamical mean-field theory. Rev Mod Phys 78:865–951
173. Ceperley D, Alder B (1986) Quantum monte carlo. Science 231:555–560
174. Wagner LK, Ceperley DM (2016) Discovering correlated fermions using quantum monte carlo. Rep Prog Phys 79:094501
175. Sun Q et al (2017) Py SCF: the python-based simulations of chemistry framework. Wiley Interdiscip Rev: Comput Mol Sci 8:e1340
176. Aspuru-Guzik A (2005) Simulated quantum computation of molecular energies. Science 309:1704–1707
177. Bravyi, S., Gambetta, J. M., Mezzacapo, A. & Temme, K. Tapering off qubits to simulate fermionic hamiltonians. Preprint at https://arxiv.org/abs/1701.08213 (2017).
178. Babbush R et al (2018) Low-depth quantum simulation of materials. Phys Rev X 8:011044
179. Kivlichan ID et al (2018) Quantum simulation of electronic structure with linear depth and connectivity. Phys Rev Lett 120:110501
180. Motta M et al (2020) Determining eigenstates and thermal states on a quantum computer using quantum imaginary time evolution. Nat Phys 16:205–210
181. Ollitrault, P. J. et al. Quantum equation of motion for computing molecular excitation energies on a noisy quantum processor. Preprint at https://arxiv.org/abs/1910.12890 (2019).
182. Cao Y et al (2019) Quantum chemistry in the age of quantum computing. Chem Rev 119:10856–10915
183. Smart SE, Mazziotti DA (2019) Quantum-classical hybrid algorithm using an error-mitigating n-representability condition to compute the mott metal-insulator transition. Phys Rev A 100:022517
184. Smart SE, Schuster DI, Mazziotti DA (2019) Experimental data from a quantum computer verifies the generalized pauli exclusion principle. Commun Phys 2:1–6
185. Bauer, B., Bravyi, S., Motta, M. & Chan, G. K.-L. Quantum algorithms for quantum chemistry and quantum materials science. Preprint at https://arxiv.org/abs/2001.03685 (2020).
186. D'Haenens-Johansson UFS et al (2011) Optical properties of the neutral silicon split-vacancy center in diamond. Phys Rev B 84:245208
187. Gali A, Maze JR (2013) Ab initio study of the split silicon-vacancy defect in diamond: electronic structure and related properties. Phys Rev B 88:235205
188. Green BL et al (2017) Neutral silicon-vacancy center in diamond: spin polarization and lifetimes. Phys Rev Lett 119:096402
189. Rose BC et al (2018) Observation of an environmentally insensitive solid-state spin defect in diamond. Science 361:60–63
190. Weber JR et al (2010) Quantum computing with defects. Proc Natl Acad Sci U S A 107:8513–8518
191. Seo H, Govoni M, Galli G (2016) Design of defect spins in piezoelectric aluminum nitride for solid-state hybrid quantum technologies. Sci Rep 6:20803
192. Seo H, Ma H, Govoni M, Galli G (2017) Designing defect-based qubit candidates in wide-gap binary semiconductors for solid-state quantum technologies. Phys Rev Mater 1:075002
193. Ivády V, Abrikosov IA, Gali A (2018) First principles calculation of spin-related quantities for point defect qubit research. npj Comput Mater 4:76

194. Dreyer CE, Alkauskas A, Lyons JL, Janotti A, Van de Walle CG (2018) First-principles calculations of point defects for quantum technologies. Annu Rev Mat Res 48:1–26
195. Anderson CP et al (2019) Electrical and optical control of single spins integrated in scalable semiconductor devices. Science 366:1225–1230
196. Abrams DS, Lloyd S (1997) Simulation of many-body fermi systems on a universal quantum computer. Phys Rev Lett 79:2586
197. Davies G, Hamer MF (1976) Optical studies of the 1.945 eV vibronic band in diamond. Proc R Soc A348:285–298
198. Rogers LJ, Armstrong S, Sellars MJ, Manson NB (2008) Infrared emission of the NV centre in diamond: Zeeman and uniaxial stress studies. N J Phys 10:103024
199. Peruzzo A et al (2014) A variational eigenvalue solver on a photonic quantum processor. Nat Commun 5:4213
200. McClean JR, Romero J, Babbush R, Aspuru-Guzik A (2016) The theory of variational hybrid quantum-classical algorithms. N J Phys 18:023023
201. Kandala A et al (2017) Hardware-efficient variational quantum eigensolver for small molecules and quantum magnets. Nature 549:242–246
202. Green BL et al (2019) Electronic structure of the neutral silicon-vacancy center in diamond. Phys Rev B 99:161112
203. Thiering G, Gali A (2019) The (eg $\otimes$ eu) $\otimes$ eg product jahn–teller effect in the neutral group-IV vacancy quantum bits in diamond. npj Comput Mater 5:18
204. Son NT et al (1999) Photoluminescence and zeeman effect in chromium-doped 4h and 6h SiC. J Appl Phys 86:4348–4353
205. Koehl WF et al (2017) Resonant optical spectroscopy and coherent control of Cr4+ spin ensembles in sic and gan. Phys Rev B 95:035207
206. Choi S, Jain M, Louie SG (2012) Mechanism for optical initialization of spin in NV-center in diamond. Phys Rev B 86:041202
207. Bockstedte M, Schütz F, Garratt T, Ivády V, Gali A (2018) Ab initio description of highly correlated states in defects for realizing quantum bits. npj Quant Mater 3:31
208. Ma H, Govoni M, Gygi F, Galli G (2018) A finite-field approach for GW calculations beyond the random phase approximation. J Chem Theory Comput 15:154–164
209. Nguyen NL, Ma H, Govoni M, Gygi F, Galli G (2019) Finite-field approach to solving the bethe-salpeter equation. Phys Rev Lett 122:237402
210. Giannozzi P et al (2009) QUANTUM ESPRESSO: a modular and open-source software project for quantum simulations of materials. J Phys Condens Matter 21:395502
211. Perdew JP, Burke K, Ernzerhof M (1996) Generalized gradient approximation made simple. Phys Rev Lett 77:3865–3868
212. Govoni M, Galli G (2015) Large scale GW calculations. J Chem Theory Comput 11:2680–2696
213. Jordan P, Wigner EP (1928) About the pauli exclusion principle. Z Phys 47:631–651
214. Kobayashi T, Salfi J, Chua C, van der Heijden J, House MG, Culcer D, Hutchison WD, Johnson BC, McCallum JC, Riemann H, Abrosimov NV, Becker P, Pohl H-J, Simmons MY, Rogge S (2020) Engineering long spin coherence times of spin–orbit qubits in silicon. Nat Mater. https://doi.org/10.1038/s41563-020-0743-3
215. Tyryshkin AM et al (2012) Electron spin coherence exceeding seconds in high-purity silicon. Nat Mater 11:143–147
216. Muhonen JT et al (2014) Storing quantum information for 30 seconds in a nanoelectronic device. Nat Nanotechnol 9:986–991
217. Veldhorst M et al (2014) An addressable quantum dot qubit with fault-tolerant control-fidelity. Nat Nanotechnol 9:981–985
218. Bar-Gill N, Pham LM, Jarmola A, Budker D, Walsworth RL (2013) Solid-state electronic spin coherence time approaching one second. Nat Commun 4:1743
219. Doherty MW, Manson NB, Delaney P, Hollenberg LCL (2011) The negatively charged nitrogen-vacancy centre in diamond: the electronic solution. N J Phys 13:025019

220. Maze JR et al (2011) Properties of nitrogen-vacancy centers in diamond: the group theoretic approach. N J Phys 13:025025
221. Doherty MW et al (2013) The nitrogen-vacancy colour centre in diamond. Phys Rep 528:1–45
222. Goldman ML et al (2015) State-selective intersystem crossing in nitrogen-vacancy centers. Phys Rev B 91:165201
223. Diler B et al (2020) Coherent control and high-fidelity readout of chromium ions in commercial silicon carbide. npj Quant Inf 6:11
224. Van der Heijden J et al (2014) Probing the spin states of a single acceptor atom. Nano Lett 14:1492–1496
225. Bir G, Butekov E, Pikus G (1963) Spin and combined resonance on acceptor centres in Ge and Si type crystals—I: paramagnetic resonance in strained and unstrained crystals. J Phys Chem Solid 24:1467–1474
226. Bir G, Butikov E, Pikus G (1963) Spin and combined resonance on acceptor centres in Ge and Si type crystals—II: the effect of the electrical field and relaxation time. J Phys Chem Solid 24:1475–1486
227. Richardson CJK, Lordi V, Misra S, Shabani J (2020) Materials science for quantum information science and technology. MRS Bull. https://doi.org/10.1557/mrs.2020.147
228. Monroe C, Kim J (2013) Scaling the ion trap quantum processor. Science 339:1164. https://www.science.org/doi/10.1126/science.1231298
229. Bar-Gill N, Pham LM, Belthangady C, Le Sage D, Cappellaro P, Maze JR, Lukin MD, Yacoby A, Walsworth R (2012) Solid-state electronic spin coherence time approaching one second. Nat Commun 3:856
230. Seidelin S, Chiaverini J, Reichle R, Bollinger JJ, Leibfried D, Britton J, Wesenberg JH, Blakestad RB, Epstein RJ, Hume DB, Itano WM, Jost JD, Langer C, Ozeri R, Shiga N, Wineland DJ (2006) Microfabricated surface-electrode ion trap for scalable quantum information processing. Phys Rev Lett 96:253003
231. Barends R, Wenner J, Lenander M, Chen Y, Bialczak RC, Kelly J, Lucero E, O'Malley P, Mariantoni M, Sank D, Wang H, White TC, Yin Y, Zhao J, Cleland AN, Martinis JM, Baselmans JJA (2011) Minimizing quasiparticle generation from stray infrared light in superconducting quantum circuits. Appl Phys Lett 99:113507. https://doi.org/10.1063/1.3638063
232. Zeng LJ, Nik S, Greibe T, Krantz P, Wilson CM, Delsing P, Olsson E (2015) Direct observation of the thickness distribution of ultra thin AlOx barrier in Al/AlOx/Al Josephson junctions. J Phys D Appl Phys 48:395308
233. De Sousa R, Whaley KB, Hecht T, Von Delft J, Wilhelm FK (2009) Microscopic model of critical current noise in Josephson-junction qubits: subgap resonances and Andreev bound states. Phys Rev B Condens Matter Mater Phys 80:094515
234. Oh S, Cicak K, Kline JS, Sillanpaa MA, Osborn KD, Whittaker JD, Simmonds RW, Pappas DP (2006) Elimination of two level fluctuators in superconducting quantum bits by an epitaxial tunnel barrier. Phys Rev B Condens Matter Mater Phys 74:100502
235. Hite DA, Colombe Y, Wilson AC, Brown KR, Warring U, Jördens R, Jost JD, McKay KS, Pappas DP, Leibfried D, Wineland DJ (2012) 100-Fold reduction of electric-field noise in an ion trap cleaned with in situ argon-ion-beam bombardment. Phys Rev Lett 109:103001
236. De Sousa R (2007) Dangling-bond spin relaxation and magnetic *1/f* noise from the amorphous-semiconductor/oxide interface: theory. Phys Rev B Condens Matter Mater Phys 76:245306
237. Lee D, DuBois JL, Lordi V (2014) Identification of the local sources of paramagnetic noise in superconducting qubit devices fabricated on α-Al_2O_3 substrates using density-functional calculations. Phys Rev Lett 112:017001. https://doi.org/10.1103/PhysRevLett.112.017001
238. Richardson CJK, Siwak NP, Hackley J, Keane ZK, Robinson JE, Arey B, Arslan I, Palmer BS (2016) Supercond Sci Technol 29:064003
239. Earnest CT, Béjanin JH, McConkey TG, Peters EA, Korinek A, Yuan H, Mariantoni M (2018) Supercond Sci Technol 31:125013

240. Frolov SM, Plissard SR, Nadj-Perge S, Kouwenhoven LP, Bakkers EPAM (2013) MRS Bull 38:809
241. Krogstrup P, Ziino NLB, Chang W, Albrecht SM, Madsen MH, Johnson E, Nygård J, Marcus CM, Jespersen TS (2015) Nat Mater 14:400
242. Fowler AG, Mariantoni M, Martinis JM, Cleland AN (2012) Phys Rev A At Mol Opt Phys 86:032324
243. Bruzewicz CD, Chiaverini J, McConnell R, Sage JM (2019) Appl Phys Rev 6:021314
244. Rosenberg D, Kim D, Das R, Yost D, Gustavsson S, Hover D, Krantz P, Melville A, Racz L, Samach GO, Weber SJ, Yan F, Yoder JL, Kerman AJ, Oliver WD (2017) NPJ Quant Inf 3:42
245. Amemiya N, Tsukamoto O (1995) IEEE Trans Appl Supercond 5:218
246. Barannikova S, Shlyakhova G, Zuev L, Malinovskiy A (2016) Int J Geomate 10:1906
247. Kang SG, Kim MG, Kim CG (2007) Compos Struct 78:440
248. Crowder T, Lanzagorta M (2019) Quantum information processing in the neighborhood of a black hole. Nat Comput. https://doi.org/10.1007/s11047-019-09737-7
249. Peres A, Scudo PF, Terno DR (2002) Quantum entropy and special relativity. Phys Rev Lett 88:230402
250. Peres A, Terno DR (2002) Relativistic Doppler effect in quantum communication. J Mod Opt 50:1165–1173
251. Céleri LC, Kiosses V, Terno DR (2016) Spin and localization of relativistic fermions and uncertainty relations. Phys Rev A 94:062115
252. Saldanha P, Vedral V (2012) Physical interpretation of the Wigner rotations and its implications for relativistic quantum information. New J Phys 14:023041
253. Taillebois ERF, Avelar AT (2013) Spin-reduced density matrices for relativistic particles. Phys Rev A 88:060302
254. Choi T (2013) Relativistic spin operator and Lorentz transformation of the spin state of a massive Dirac particle. J Korean Phys Soc 62:1085–1092
255. Bauke H et al (2014) Relativistic spin operators in various electromagnetic environments. Phys Rev A 89:052101
256. Martin K (2008) The scope of a quantum channel. Proc Symp Appl Math 71:183–211
257. Lanzagorta M (2013) Quantum information in gravitational fields. Institute of Physics, San Rafael, CA
258. Alsing PM, Stephenson GJ, Kilian P (2009) Spin-induced non-geodesic motion, gyroscopic precession, Wigner rotation and EPR correlations of massive spin 1/2 particles in a gravitational field. arXiv:0902.1396v1 [quant-ph]
259. Terashima P, Ueda M (2004) Einstein–Rosen correlation in gravitational field. Phys Rev A 69:032113
260. Crowder T, Lanzagorta M (2018) The scope of a relativistic quantum process with spin-momentum entanglement. In: Stepney S, Verlan S (eds) Unconventional computation and natural computation. Lecture Notes in Computer Science, vol 10867. UCNC, 2018
261. Irkhin VY, Skryabin YN (2019) Modern physics of the condensed state: strong correlations and quantum topology. Phys Met Metallogr. https://doi.org/10.1134/s0031918x19060061
262. Anderson PW (1987) The resonating valence bond state in La_2CuO_4 and superconductivity. Science 235:1196–1198
263. Anderson PW (2011) Personal history of my engagement with cuprate superconductivity, 1986–2010. Int J Mod Phys B 25:1–39
264. Wen XG (2004) Quantum field theory of many-body systems—from the origin of sound to an origin of light and electrons. Oxford University Press
265. Levin MA, Wen XG (2005) Photons and electrons as emergent phenomena, 2005. Rev Mod Phys 77:871–880
266. Sachdev S (2011) Quantum Phase Transitions. Harvard University, Massachusetts
267. Coleman P (2003) Many body physics: unfinished revolution. Ann Henri Poincaré 4:559–580
268. Witten E (1989) Quantum field theory and the Jones polynomial. Commun Math Phys 121:351–399

269. Zeng B, Wen XG (2015) Gapped quantum liquids and topological order, stochastic local transformations and emergence of unitarity. Phys Rev B 91:125121
270. Swingle B, McGreevy J (2016) Renormalization group constructions of topological quantum liquids and beyond. Phys Rev B 93:045127
271. Irkhin VY, Katsnelson MI (1990) Ground state and electron-magnon interaction in an itinerant-electron ferromagnet: half-metallic ferromagnets. J Phys Condens Matter 2:7151–7171
272. Anderson PW, Casey PA (2009) Transport anomalies of the strange metal: resolution by hidden Fermi liquid theory. Phys Rev B 80:094508
273. Irkhin VY (2017) Unusual magnetism of the Kondo lattice. Phys–Usp 60:747–761
274. Mermin ND (1979) The topological theory of defects in ordered media. Rev Mod Phys 51:591–648
275. Hertz JA (1976) Quantum critical phenomena. Phys Rev B 14:1165–1184
276. Millis AJ (1993) Effect of nonzero temperature on quantum critical points in itinerant fermion systems. Phys Rev B 48:7183–7196
277. Anderson PW (2008) Hidden Fermi liquid: the secret of high-Tc cuprates. Phys Rev B 78:174505
278. Zeng B, Chen X, Zhou D-L, Wen X-G (2015) Quantum information meets quantum matter, from quantum entanglement to topological phase in many-body systems, In the Springer Book Series—Quantum Information Science and Technology (in press), arXiv preprint arXiv:1508.02595.
279. Hartnoll SA, Lucas A, Sachdev S (2018) Holographic quantum matter. MIT Press, Cambridge, MA. ArXiv:1612.07324
280. Sachdev S (2012) What can gauge-gravity duality teach us about condensed matter physics? Annu Rev Condens Matter Phys 3:9. ArXiv:1108.1197
281. Garhwal S, Ghorani M, Ahmad A (2019) Quantum programming language: a systematic review of research topic and top cited languages. Archiv Computat Meth Eng. https://doi.org/10.1007/s11831-019-09372-6
282. Shor PW (1994) Algorithms for quantum computation: discrete logarithms and factoring. In: Proceedings of 35th annual symposium on the foundations of computer science, 20–22 Nov 1994, Santa Fe, NM, USA, pp 124–134
283. Unruh D (2006) Quantum programming languages. Informatik-Forschung and Entwicklung 21(1):55–63
284. Omer B (2003) Structured quantum programming. PhD dissertation, TU Vienna. http://tph.tuwien.ac.at/~oemer/doc/structquprog.pdf. Accessed 8 Apr 2019
285. Mlnarik H (2006) Introduction to LanQ: an imperative quantum programming language. http://lanq.sourceforge.net/doc/introToLanQ.pdf. Accessed 8 Apr 2019
286. Mlnarik H (2007) Quantum programming language LanQ. PhD thesis, Masaryk University
287. Pakin S (2016) A quantum macro assembler. In: IEEE high performance extreme computing conference (HPEC), 13–15 Sept 2016, Waltham, MA, USA. https://doi.org/10.1109/HPEC.2016.7761637
288. Selinger P (2004) Towards a semantics for higher-order quantum computation. In: Proceeding of 2nd international workshop on quantum programming languages, Turku, Finland, 12–13 July 2004, pp 127–143
289. Selinger P (2004) Towards a quantum programming language. Math Struct Comput Sci 14(4):527–586
290. Chakraborty A (2011) QuECT: a new quantum programming paradigm. arXiv:1104.0497
291. Killoran N, Izaac J, Quesada N, Bergholm V, Amy M, Weedbrook C (2018) Strawberry fields: a software platform for photonic quantum computing. arXiv:1804.03159
292. Krämer S, Plankensteiner D, Ostermann L, Ritsch H (2018) QuantumOptics.jl: a Julia framework for simulating open quantum systems. Comput Phys Commun 227:109–116
293. Q# Language: https://www.microsoft.com/en-us/quantum/development-kit. Accessed 8 Apr 2019
294. Feitosa SS, Vizzotto JK, Piveta EK, Bois ARD (2016) FJQuantum: a quantum object oriented language. Electron Notes Theor Comput Sci 324:67–77

295. Lacerda FG, Renes JM, Renner R (2019) Classical leakage resilience from fault-tolerant quantum computation. J Cryptol. https://doi.org/10.1007/s00145-019-09310-6
296. Aliferis P, Gottesman D, Preskill J (2006) Quantum accuracy threshold for concatenated distance-3 codes. Quant Inform Comput 6(2):97–165
297. Ghasemian E (2023) Stationary states of a dissipative two-qubit quantum channel and their applications for quantum machine learning. Quant Machine Intell. https://doi.org/10.1007/s42484-023-00096-2
298. Schuld M, Sinayskiy I, Petruccione F (2014) The quest for a quantum neural network. Quant Inf Process 13:2567
299. Rabinovich MI, Varona P, Selverston AI, Abarbanel HD (2006) Dynamical principles in neuroscience. Rev Mod Phys 78:1213
300. Ji Y-H, Liu Y-M (2013) Investigations into quantum correlation of coupled qubits in a squeezed vacuum reservoir. Chin Phys B 22(2):020305
301. Kowalewska-Kudłaszyk A, Leoński W (2010) Squeezed vacuum reservoir effect for entanglement decay in the nonlinear quantum scissor system. J Phys B Atom Mol Opt Phys 43:205503
302. Ghasemian E, Tavassoly M (2017) Quantum dynamics of a BEC interacting with a single-mode quantized eld under the inuence of a dissipation process: thermal and squeezed vacuum reservoirs. Laser Phys 27:095202
303. Khrennikov A (2020) Quantum versus classical entanglement: eliminating the issue of quantum nonlocality. Found Phys 50(12):1762–1780
304. Duan L-M, Kimble H (2003) Efficient engineering of multiatom entanglement through single-photon detections. Phys Rev Lett 90:253601
305. Zhao B, Chen Z-B, Chen Y-A, Schmiedmayer J, Pan J-W (2007) Robust creation of entanglement between remote memory qubits. Phys Rev Lett 98(24):240502

Chapter 2
Quantum Programming

Introduction by the Editor

Let us begin this chapter with hybrid quantum algorithms and hybrid runtimes, workflows to integrate hybrid quantum applications, and source of origin to monitor and analyze the workflows. Quantum programming resolves the security risk assessment, anomalies, and management strategy. Subsequently, this chapter discusses an open-source package for implementing parameterized quantum circuits used in variational quantum algorithms (QVAs) and quantum machine learning (QML) algorithms. Quantum programming proposes as well motivates the researchers to use XOR-And-Inverter (XAG) for executing quantum compilation. In continuation to quantum computing, network security intrusion detection has been explored using quantum machine learning methods. It is proved that the QML methods are applied to analyze on a high traffic data streaming network in the cloud or server environment.

In this chapter, we describe a method to bridge or coordinate the existing conceptual gap between workflow-based orchestration of hybrid quantum-classical applications and the efficiency improvements achieved using hybrid runtimes. Apparently, Quantum applications typically involve a mix or combination of quantum and classical programs that need to be orchestrated efficiently as shown in this book. Thereby we conclude "Quantum applications = quantum program + classical program" stated in this chapter. Just merely workflows have been used by us for orchestration, they may not be the most efficient solution for certain quantum algorithms that require iterative execution of quantum and classical programs. At a large sense, QML has been successful in our modern data science activities.

Next step into Information Security Management Systems (ISMS). However, we provide a solution to the incident response problem within Information Security Management Systems (ISMS) using quantum computing. Indeed, ISMS minimizes an entity's risk and supports performance continuity plan. The incidents or risks

V. Natarajan (ed.), *Quantum Artificial Intelligence*,
https://doi.org/10.1007/978-981-96-5051-4_2

vary in severity, and each level of the incident requires a specific set of mitigation controls to restore the ISMS. Since time is a critical factor or priority at an incident response, the classic solutions become inefficient as the number of incidents increases, especially in scenarios where security management is offered as a service to multiple companies. Therefore, Quantum computing offers the advantage of providing solutions in a constant time, regardless of the number of incidents being handled. An open-source Python package qLEET has been designed for studying parameterized quantum circuits (PQCs) commonly employed in variational quantum algorithms (VQAs) and quantum machine learning (QML) algorithms. Both VQAs and QML could help to resolve some unsolved hard problems in data analytics. The qLEET package facilitates the depth analysis of PQCs by enabling computations of properties such as expressibility and entangling power. Through research studies on the entanglement spectrum and the distribution of parameterized states generated by the PQC, the process will enhance the communication networks.

Let us introduce the usage of Xor-And-Inverter Graphs (XAG) as a means of specifying Boolean arithmetic functions for quantum compilation. There are different algorithms (mentioned in this chapter) based on XAGs to synthesize quantum circuits in the Clifford + T library, with the aim of targeting fault-tolerant quantum computing. These algorithms serve to reduce the number of qubits, the T-count, and the T-depth order to reduce huge quantum bits during execution, the quantum research proposes quantum systems using XAGs to perform Boolean arithmetic operations for quantum compilation. A structured and efficient representation of Boolean arithmetic operations, XAGs facilitate. It could be leveraged for optimizing quantum circuits. Next stage, we explore and exploit the application of quantum machine learning (QML) methods to resolve the challenges posted by big data technology in the domain of intrusion detection systems. There are limitations of classical methods handling volume of data, Quantum machine learning methods offer potential solutions to overcome the problems when dealing with big data analytics. So far, no proper prediction models accurately handle outliers. But QML handles the outliers in a better way. Quantum support vector machine (QSVM) and quantum convolutional neural network (QCNN) are discussed in this book as promising methods for intrusion detection, leveraging the computational power of quantum hardware. It introduces a novel method for analyzing network traffic using quantum machine learning (QML). In our research world, Game theory and Multi-dimensional linear programming handle network optimization problems whereas QML and Quantum Statistics perform well on optimization. Still practical implementation and regress testing on progress for optimization. For encoding network traffic data into terms suitable for quantum computation and leveraging quantum algorithms for network attack detection, QML is aptly suitable to use. To encode network traffic data into terms that are compatible with quantum computing, QML system handles in a better way. The efficient quantum algorithms can effectively process data encoding techniques over the classical methods through research and industry experience. The experimental results showcase the effectiveness of the proposed QML-derived approach on detecting network attacks accurately. By comparing the performance of QML with traditional machine learning and deep learning methods, the research and development emphasize the advantages of quantum computing in the context of network traffic analysis and optimization over neural networks and Game theory.

For efficiently mapping quantum computational resources using hierarchical assembly code, we introduce a novel approach to support large-scale quantum computing in order to increase memory and space utilization. This is an advance development in Quantum computing facilitating a fast-mapping process by leveraging hierarchical assembly code. In addition to this, while comparing to non-structured code, the quantum computing factors or cracks a 512-bit integer using the Shor algorithm from a long duration (1500 days) to just 1 hour through enabling quicker mapping for Crypto systems, significantly bringing down the performance time required for Information security tasks. The Shor algorithm applies to crack the integer factorization through lesser framebuffers and swap chains. Hence the known RSA-1024 crypto algorithm will be cracked by quantum computing in a matter of time.

At the end of Chap. 2, we present the Hybrid system for Quantum Lambda Calculus (QLC) along with the development of a linear logical framework, specifically focusing on a practical version called Proto-Quipper. The QLC solves numerical approximation problems faster than the classical Calculus. Further, the QLC is a formal algebraic framework system for reasoning and computation. It serves a computational framework equivalent to the Turing machine. Apart from quantum computing, Quipper is a fast and functional quantum programming language that has gained popularity in the quantum computing community. In reality, the quantum research studies the Hybrid system with a linear specification logic (SL) to reason about the linear type of system of Quipper. This logic allows for the formalization of typing and evaluation rules, enabling rigorous reasoning about the semantics of Proto-Quipper. Due to the advancement of QLC, a lot of quantum algorithms will come out soon. That could help and ease semantics issues for the nonlinear type of system of Quipper.

Machine Generated Summaries

Disclaimer: The summaries in this chapter were generated from Springer Nature publications using extractive AI auto-summarization: An extraction-based summarizer aims to identify the most important sentences of a text using an algorithm and uses those original sentences to create the auto-summary (unlike generative AI). As the constituted sentences are machine selected, they may not fully reflect the body of the work, so we strongly advise that the original content is read and cited. The auto generated summaries were curated by the editor to meet Springer Nature publication standards. To cite this content, please refer to the original papers.

Machine generated keywords: workflow, logic, intrusion, program, malicious, stream, hybrid, risk, security, visualize, mapping, quantum algorithm, library, support, attack.

Provenance-Preserving Analysis and Rewrite of Quantum Workflows for Hybrid Quantum Algorithms [1]

This is a machine-generated summary of:

Weder, Benjamin; Barzen, Johanna; Beisel, Martin; Leymann, Frank: Provenance-Preserving Analysis and Rewrite of Quantum Workflows for Hybrid Quantum Algorithms [1].

Published in: SN Computer Science (2023).

Link to original: https://doi.org/10.1007/s42979-022-01625-9

If you want to cite the papers, please refer to the original.

For technical reasons we could not place the page where the original quote is coming from.

Abstract-Summary

"Quantum applications are hybrid, i.e., they comprise quantum and classical programs, which must be orchestrated."

"The orchestration using workflows can be inefficient for some quantum algorithms, requiring the execution of quantum and classical programs in a loop."

"The quantum and classical programs are combined in a single hybrid program, for which the execution is optimized."

"This leads to a conceptual gap between the modeling benefits of workflow technologies, e.g., modularization, reuse, and understandability, and the efficiency improvements when using hybrid runtimes."

"To close this gap, we introduce a method to model all tasks explicitly in the workflow model and analyze the workflow to detect parts of the workflow that can benefit from hybrid runtimes."

"Corresponding hybrid programs are automatically generated based on the quantum and classical programs, and the workflow is rewritten to invoke them."

Introduction

"By providing various advantages, such as scalability, reliability, robustness, automated error handling, or transactional processing, workflows are a promising means to orchestrate the quantum and classical programs of a hybrid quantum application [2–4]."

"Although workflows provide a variety of benefits when modeling hybrid quantum applications, the orchestration can get inefficient for some use cases."

"We present a method to benefit from the advantages of modeling hybrid quantum applications using workflows while increasing the efficiency by utilizing hybrid runtimes for appropriate workflow parts."

"To overcome this issue, we automatically collect provenance data about the running hybrid programs, such as the current state or intermediate results, which are used to provide process views to the user, allowing to visualize the original and the rewritten workflow with its current state in the workflow engine."

Fundamentals and Problem Statement

"The required background about hybrid quantum algorithms and hybrid runtimes, workflows to orchestrate hybrid quantum applications, and provenance to monitor and analyze the workflows are introduced."

"Different hybrid algorithms execute the quantum and classical programs interleaved multiple times, e.g., variational algorithms perform classical optimization and quantum program execution in a loop [5, 6]."

"The hybrid runtime optimizes the data transfer time by closely deploying the classical programs to the used quantum computer."

"Due to the heterogeneity of the quantum and classical programs implementing a hybrid algorithm, the orchestration of the control and data flow between them can be challenging [7, 8]."

"QuantME provides explicit modeling constructs for different frequently occurring pre-processing, quantum program execution, and post-processing tasks."

"Orchestrating the interleaved execution of quantum and classical programs using workflows is inefficient due to the queuing and data transfer times."

Quantum Workflow Analysis and Rewrite

"If the workflow part does not execute quantum circuits, a hybrid runtime is not needed, and the deployment, e.g., in the cloud is sufficient [3]. In addition to the execution of quantum circuits, the workflow part must also orchestrate the execution of classical programs or services."

"This has to be supported by the hybrid runtime, and the corresponding hybrid program must be generated accordingly if the candidate comprises such an event [5].

Gateways: Similar to events, gateways can be difficult to realize within hybrid runtimes if no workflow engine is used internally."

"If different hybrid runtimes were selected for the candidates of the workflow model in step 3, multiple provenance collectors must be generated."

"The generated hybrid programs and provenance collectors must be deployed to the hybrid runtimes before the workflow can be executed."

Monitoring and Analysis of Rewritten Quantum Workflows

"Further, it is shown how this data can be used to instrument process views for quantum workflows."

"To monitor and analyze hybrid programs executed in a hybrid runtime and to support the usage of process views for quantum workflows, detailed provenance data must be collected."

"The right side shows a process view of the workflow before rewriting, orchestrating a hybrid loop with two gateways and three tasks."

"Visualizing the process view, the token moves between the different tasks of the hybrid loop, enabling the user to monitor the progress."

"While in the executed workflow model only the input parameters and the output after finishing the execution are available, the process view provides further information, e.g., the current iteration, the counts from the last ansatz execution, or the current costs."

Prototypical Validation

"The QuantME Transformation Framework is a graphical BPMN workflow modeler based on the Camunda modeler [9] that was extended to support QuantME."

"It extracts the quantum and classical programs for each candidate and sends them to the Hybrid Runtime Handler supporting the selected hybrid runtime."

"The hybrid program and the provenance collector are sent back to the Workflow Rewriter."

"Further, the QuantME Transformation Framework is extended by a View Generator to create different process views during workflow rewrite and transformation, which can be selected by the user for monitoring and analysis."

"It was extended to assist the user during workflow analysis and rewrite, e.g., visualizing candidates and enabling the selection of a hybrid runtime if multiple are suitable."

"Further, we prototypically implemented two Hybrid Runtime Handlers, namely the Qiskit Runtime Handler and the Amazon Braket Handler."

"For the execution of generated hybrid programs, Qiskit Runtime [10] and Amazon Braket Hybrid Jobs [11] are used."

Case Study

"After receiving the pre-processed input data, clustering is performed using the quantum k-means algorithm [12], consisting of multiple tasks that need to be orchestrated."

"The quantum algorithm is initialized by determining random initial centroids for the clustering and based on these centroids corresponding quantum circuits are generated."

"A hybrid loop is entered using the quantum circuit execution task to execute the generated quantum circuits."

"Similar to the clustering, it then enters a hybrid loop executing the ansatz and optimizing the parameters in each iteration based on the measurement results."

"The last user task provides details about the performed clustering and the trained classifier, which can be analyzed."

"The two service tasks initializing the hybrid algorithms and the user tasks are unchanged."

"The optimization candidates of the original workflow are both replaced by a service task invoking the generated hybrid program at Qiskit Runtime."

Evaluation

"We compare the execution times of various workflows without using our method to the execution times after rewriting and using hybrid runtimes from AWS and IBM."

"We focus on three attributes characterizing the workflows, directly influencing the execution time of the analysis and rewrite method: (1) the number of activities within the workflow model enlarges the search space for the detection of candidates."

"This workflow model was evaluated for both providers, i.e., IBM and AWS, to determine if this leads to differences in the program generation and rewriting times."

"As the number and size of the candidates are the same, the program generation and workflow rewrite times are almost equal to workflow 1."

"The other workflows do not increase the overall number of activities but add additional candidates or adapt their size to specifically evaluate the program generation and rewrite."

Discussion

"Such runtimes can incorporate a workflow engine or an HPC, which are co-located to the quantum computers."

"For the candidate detection and filtering, only the structure of the workflows, i.e., the tasks, events, gateways, and the corresponding quantum and classical programs, as well as the characteristics of the hybrid runtimes, are considered at the moment."

"Including non-functional requirements in the detection and filtering, as well as the selection of a concrete hybrid runtime to use, can improve the resulting rewritten workflow [13–15]."

"To improve the efficiency, tasks decreasing the size should be performed before transmitting the data between the hybrid runtime and the workflow engine [16]."

"The quantum circuit execution tasks allow defining a provider or even a specific quantum computer to use, which also restricts the set of possible hybrid runtimes for the rewrite."

Related Work

"Bucchiarone et al. [17] introduce an approach to model workflows abstractly, where the tasks are annotated with goals, preconditions, and effects."

"Mundbrod et al. [18] follow a similar approach, adapting workflows by injecting workflow fragments depending on the current context, e.g., the available resources or the load on a component."

"Képes et al. [19] adapt workflow models by dynamically selecting and executing suitable workflow fragments based on the current situation."

"Various approaches are generating new workflow models or adapting existing ones using the data collected during workflow execution in the audit trail of the workflow engine by applying process mining techniques [20–22]."

"To rewriting scientific workflows, Cohen-Boulakia et al. [23] also show how to use process views to analyze collected provenance data."

"Schumm et al. [24] utilize process views for ensuring the compliance of workflow executions."

Conclusion and Future Work

"Quantum workflows enable orchestrating the control and data flow between quantum and classical programs of hybrid quantum applications."

"When executing the quantum and classical programs interleaved multiple times, the orchestration using workflows is inefficient due to the increased latency and queuing times."

"For such use cases, hybrid runtimes are provided, optimizing the execution by closely deploying the quantum and classical programs together and reducing the queuing times."

"We presented a method to automatically detect workflow parts that can benefit from hybrid runtimes."

"Further, for our experiments, the speed-up when using hybrid runtimes was about factor 2 compared to the execution of the original workflow directly accessing the quantum computers."

“We plan to investigate how to determine if other workflow parts comprising the interleaved execution of quantum and classical programs without a loop can benefit from hybrid runtimes and what factors must be considered.”

Minimizing Incident Response Time in Real-World Scenarios Using Quantum Computing [25]

This is a machine-generated summary of:

Serrano, Manuel A.; Sánchez, Luis E.; Santos-Olmo, Antonio; García-Rosado, David; Blanco, Carlos; Barletta, Vita Santa; Caivano, Danilo; Fernández-Medina, Eduardo: Minimizing incident response time in real-world scenarios using quantum computing [25].

Published in: Software Quality Journal (2023).

Link to original: https://doi.org/10.1007/s11219-023-09632-6

If you want to cite the papers, please refer to the original.

For technical reasons we could not place the page where the original quote is coming from.

Abstract-Summary

“Each incident has associated a level of severity and a set of mitigation controls, so in order to restore the ISMS, the appropriate set of controls to mitigate their damage must be selected.”

“Classic solutions are efficient in resolving scenarios with a moderate number of incidents in a reasonable time, but the response time increases exponentially as the number of incidents increases.”

“This makes classical solutions unsuitable for real scenarios in which a large number of incidents are handled and even less appropriate for scenarios in which security management is offered as a service to several companies.”

"This paper proposes a solution to the incident response problem that acts in a minimal amount of time for real scenarios in which a large number of incidents are handled."

"It applies quantum computing, as a novel approach that is being successfully applied to real problems, which allows us to obtain solutions in a constant time regardless of the number of incidents handled."

Introduction

"A key element of each ISMS is the security risk assessment and management strategy [26–29], but security risks do not only affect to ICT components of organizations, but also their business processes, and even the organization and strategy level [30]."

"Of all, most security incidents are caused by the general lack of awareness of risk or their inaccurate assessment [31]."

"The aim of this paper is to try to contribute to the resolution of a specific problem of security incident response, which is a fundamental aspect of ISMS and in particular is a part of risk management, and which is responsible for reacting to incidents by applying controls to reduce damage and efficiently restore systems [32]."

"The problem proposed in this paper concerns the optimization of incident response in a risk analysis and management system, where incident response can be optimized by selecting those appropriate controls to perform, being a problem that grows exponentially with the number of incidents."

Background and Related Work

"This section includes background content about the three research topics addressed in this paper, quantum programming, quantum optimization and security incident response."

"To have real applications of quantum computing, programming languages are needed that provide structured and high-level descriptions of quantum algorithms, without reference to the underlying hardware [33]."

"The main imperative quantum programming languages are Q# [34], Q|SI⟩ [35], ProjectQ [36] and Qiskit [37], among others."

"The philosophy of quantum programming is oriented toward exploring and searching optimal solutions in a probabilistic space [38]."

"This quantum computing approach will specify incidents with their associated threats and controls and search for the minimum energy state that represents the best solution for incident resolution in the shortest possible time."

"Above all, utilities and processes provide them with mechanisms that facilitate decision-making to optimize the selection and prioritization of security incidents to be resolved [39]."

MARISMA Framework for Managing Security Risks and Incidents

"This process is fully implemented by eMARISMA, which provides a workflow to (i) enter the security incident information (a description, the cause, the responsible person, and the time limits to be solved), (ii) select from the stored information and according to the data relationships defined by the risk pattern the hierarchy of elements that are involved with the security incident (threats, assets and controls), defining other related information such as the severity of the incident, and quarantine the affected controls by temporarily lowering their coverage level while the incident is resolved, and finally, once the incident is solved, (iii) support knowledge management and learning from the security incidents occurred by recording the lesson learned, incident resolution costs and some concluding remarks."

Quantum Algorithm for Incident Response Optimization

"To solve the problem, we should select a result in which the minimum set of incidents to be solved is selected, so that we cover all the controls that allow us to solve the other incidents."

"In order to solve the problem, we will model this problem as a Quadratic Unconstrained Binary Optimization (QUBO) problem, also known as unconstrained binary quadratic programming (UBQP), which will represent the objectives and constraints of our problem and can be sent to the solver of the adiabatic quantum computer to find the minimum energy state, which will coincide with the combination of variables, i.e., incidents, that must be selected to find an optimal result to our problem."

"In the light of these results we can consider it appropriate to believe that the adiabatic quantum approach for solving optimization problems in the context of security incident management is widely efficient and an improvement over previous management based on classical optimization algorithms."

Conclusions

"Security management, risk analysis and, in particular, risk management based on the correct management and learning from security incidents have become increasingly important."

"The response time offered by classic solutions grows exponentially as the number of incidents increases, making them unsuitable for real-world scenarios."

"We can state that although today there are numerous open problems related to security incident management, especially when dealing with large volumes of data, some of them can be solved using quantum algorithms."

"Part of our future work is to further investigate quantum algorithms and swarm intelligence applied to the exploitation of our security dataset of security risks and incidents from many organizations, in order to correlate security incidents in real time, providing a global and much more efficient way of responding against security incidents."

qLEET: Visualizing Loss Landscapes, Expressibility, Entangling Power and Training Trajectories for Parameterized Quantum Circuits [40]

This is a machine-generated summary of:

Azad, Utkarsh; Sinha, Animesh: qLEET: visualizing loss landscapes, expressibility, entangling power and training trajectories for parameterized quantum circuits [40].

Published in: Quantum Information Processing (2023).

Link to original: https://doi.org/10.1007/s11128-023-03998-z

If you want to cite the papers, please refer to the original.

For technical reasons we could not place the page where the original quote is coming from.

Abstract-Summary

"QLEET, an open-source Python package for studying parameterized quantum circuits (PQCs), which are widely used in various variational quantum algorithms (VQAs) and quantum machine learning (QML) algorithms."

"qLEET enables computation of properties such as expressibility and entangling power of a PQC by studying its entanglement spectrum and the distribution of parameterized states produced by it."

"In our work, we demonstrate how qLEET provides opportunities to design and improve hybrid quantum-classical algorithms by utilizing intuitive insights from the ansatz capability and structure of the loss landscape."

Introduction

"It becomes imperative to design optimal PQCs for a given problem."

"This is not straightforward because their design depends not only on the problem instances themselves but also on the quantum hardware that executes them."

"There exist three main classes of ansätze: (i) problem-inspired ansatz, where the evolutions of generators derived from properties of the given system are used to construct the PQCs [41], (ii) hardware-efficient ansatz, where a minimal set of quantum gates native to a given device are used to construct the PQCs [42], and (iii) adaptive ansatz, which is midway between the former two ansätze [43]."

"Using these three classes, one can develop numerous ansatz designs for any given problem."

"The primary motivation behind the development of qLEET stems from this need to have a framework for analyzing the capabilities of parameterized quantum circuits and comparing their performances."

"It does so by allowing users to study various properties related to the behavior of PQCs and assess their effectiveness for a given problem instance."

Overview

"All the functionalities present in qLEET are grouped under four modules, which reside under the top-level module called qleet."

"It also contains CircuitDescriptor, which allows for the building of PQC using any supported framework, therefore making the computation software agnostic, and MetaLogger, which maintains the record for events that happen during qLEET's execution."

"Simulators module: qleet.simulator contains the simulation engine for performing the computation."

"Analyzers module: qleet.analyzers performs execution of CircuitDescriptor object using PQCTrainer or CircuitSimulator functions present in the qleet.simulator module."

"Qleet.analyzers acts as a linkage between the previous two modules and is responsible for estimating various essential properties regarding PQC."

Features

"It would be much easier for a descent-based optimizer to traverse to global minima in case of higher p. This and similar loss landscape calculations in qLEET are done using the loss_landscape function present in the analyzer module."

"All such methods provide beneficial insights about the structure of the loss landscape using which one could adapt their training strategy by tweaking the optimization routine, evaluation metric and so on. In many cases, just looking at the loss landscape for a given PQC model is not enough as we define the subspace S using two of many possible directions as axes by taking linear combinations of variational parameters, while the loss landscape itself is highly nonlinear."

"Similar to the case of loss landscape visualization, each of the mentioned techniques reveals different trajectory characteristics depending on its ability to preserve both global and local structures of higher-dimensional data in low dimensional subspace."

Challenges

"We will discuss some key challenges that we come across in variational quantum computation and possible ways to identify and mitigate these problems by using tools provided in qLEET."

"In order to realistically simulate and characterize the performance of a parameterized quantum circuit (PQC), we must include these errors in our computation."

"Another source of error in quantum computation arises from the limited number of times the circuit is repeatedly executed for sampling."

"In qLEET, the default value of the number of repetitions is 1024 and is determined by the shots variable, which can be provided at the time of calling any analysis function from the analyzer module."

"In qLEET, one can potentially visualize the BP phenomena by visualizing the loss landscape for a chosen PQC and cost function."

"This could allow users to see if BP can be mitigated by tweaking either the structure of PQC itself or just the cost function."

Conclusion

"This paper presents an open-source library called qLEET and demonstrates its ability to analyze various properties of parameterized quantum circuits (PQCs), such as their expressibility and entangling power."

"We have discussed and showed how important insights could be gained from visualizing loss landscapes and training trajectories for variational quantum computation."

"We demonstrate how different modules included in qleet can be used by users to study various variational algorithms and quantum machine learning models."

"We conclude that qLEET will provide opportunities for the quantum community to design new hybrid algorithms by utilizing intuitive insights from the ansatz capability and structure of the loss landscape."

Xor-And-Inverter Graphs for Quantum Compilation [44]

This is a machine-generated summary of:

Meuli, Giulia; Soeken, Mathias; De Micheli, Giovanni: Xor-And-Inverter Graphs for Quantum Compilation [44].

Published in: npj Quantum Information (2022).

Link to original: https://doi.org/10.1038/s41534-021-00514-y

If you want to cite the papers, please refer to the original.

For technical reasons we could not place the page where the original quote is coming from.

Abstract-Summary

"We propose and motivate the use of Xor-And-Inverter Graphs (XAG) to specify Boolean functions for quantum compilation."

"Three different XAG-based compilation algorithms to synthesize quantum circuits in the Clifford + T library, hence targeting fault-tolerant quantum computing."

"The algorithms are designed to minimize relevant cost functions, such as the number of qubits, the T-count, and the T-depth, while allowing the flexibility of exploring different solutions."

Introduction

"As we target fault-tolerant quantum computing, we compile into the Clifford + T universal library and focus on the following cost functions: the T-count—the number of generated T gates; the T-depth—the maximum number of T gates to be performed sequentially, also referred to as number of T-stages; and the number of qubits."

"We propose XAG-based compilation as the method of choice to automatically synthesize quantum circuits implementing cryptographic and arithmetic logic functions with application in post-quantum cryptography and fault-tolerant quantum computing."

"An XAG logic network representing an n-variable Boolean function with inputs $x_1, \ldots, x_n$ is modeled as a Boolean chain with stepsfor $n < i \leq n + r$, depending on whether the step computes the 2-input XOR or the 2-input AND operation, where r is the number of steps."

"To providing a very compact representation for Boolean functions, XAG networks have another characteristic that makes them excellent data structures for quantum compilation: each node represents a logic function for which a convenient quantum circuit implementation exists."

Results

"While Algorithm 2 minimizes the T-depth without increasing the number of T gates, but relying on an increased number of additional qubits."

"The number of T gates achieved is equal to 4 times the multiplicative complexity of the network for both algorithms."

"The second algorithm obtains a T-depth equal to the multiplicative depth of the network."

"The comparison shows a significant reduction in both T-count and T-depth, while facing a less significant increase in number of qubits."

"That the authors of ref. [45] only report the number of Toffoli gates and the Toffoli-depth."

"We obtain the corresponding T-count and T-depth by considering the Clifford+T implementation of the Toffoli gate with 7 T gates and a T-depth equal to 3, which is optimal [46]."

"For every fixed number of qubits we report two points: the non-optimized and the optimized results."

Discussion

"Algorithm 2 achieves a T-depth equal to the multiplicative depth; the qubit overhead to achieve such T-depth depends on the number of shared inputs in the linear transitive fan-ins of the AND nodes in a level."

"If two AND nodes share the same input signal, the corresponding quantum circuit will have a T-depth equal to 2, as each AND implementation will add a T gate to the shared qubit."

"In order to achieve this result, our second algorithm copies inputs that are shared among more AND nodes in a level on new qubits."

"The compilation will request a new qubit whenever inputs are shared among AND nodes at the same level in the XAG."

"If we sum the number of AND nodes in a level with the number of shared inputs among them, we obtain a quantity equal to the number of helper qubits required to compile that level."

Methods

"Since the algorithm dedicates one helper qubit for each node of the XAG to store its computed Boolean function, we use nodes' identifiers, e.g. x_i, as parameters for quantum operations, e.g., NOT(x_i), meaning that the operation is performed on the corresponding qubits."

"If the node shares some inputs with another, a new qubit will be assigned to compute the corresponding parity function, otherwise a qubit corresponding to a node in the fan-in cone is used."

"The move of placing a pebble on a node corresponds to computing the logic of that node on this helper qubit."

"Once a strategy for pebbling the abstract graph is found, each time a pebble is placed on a box node which compresses x_i the compute (x_i) function will be called, while whenever a pebble is removed from a node, the compute†(x_i) function will be called to uncompute the node."

Security Intrusion Detection Using Quantum Machine Learning Techniques [47]

This is a machine-generated summary of:

Kalinin, Maxim; Krundyshev, Vasiliy: Security intrusion detection using quantum machine learning techniques [47].

Published in: Journal of Computer Virology and Hacking Techniques (2022).

Link to original: https://doi.org/10.1007/s11416-022-00435-0

If you want to cite the papers, please refer to the original.

For technical reasons we could not place the page where the original quote is coming from.

Abstract-Summary

"The paper observes the quantum machine learning (QML) methods overcoming the barriers of big data and the computing abilities of common hardware for the purpose of high performance intrusion detection."

"Quantum support vector machine (QSVM) and quantum convolution neural network (QCNN) as concurrent methods are discussed and evaluated comparing to the conventional intrusion detectors running on the traditional computer."

"Experimental results show the ability of the QML-based intrusion detection for processing big data inputs with high accuracy (98%) providing a twice faster speed comparing to the conventional machine learning algorithms utilized for the same task."

Introduction

"In case of huge data inputs, traditional machine learning (ML) methods meet several barriers: a wide variety of possible intrusions forces us to create and handle extremely big databases of the malicious samples."

"Any ML-based classification done on a big data input significantly degrades the IDS's performance on the training and testing stages [48, 49]; new intrusions specific to smart networks (e.g., the forced power consumption, the dynamic topology re-building) have a wide class of polymorphic attacks that have mutations, namely, local differences, omissions, blank spaces in the operational sequences."

"The major goal of such works is to reduce the dependence of the ML methods on the long search at large-size databases of the malicious signatures, enrich an ability of the ML models to self-training at the attack identification, identify unknown and polymorphic intrusions, and attain the higher performance of the intrusion detection."

"We focus on the novel approach of the quantum machine learning applied to the intrusion detection, specifically, in case of big inputs."

The Quantum Machine Learning

"The main idea of the quantum information theory is that a large class of problems that require resources and time exponentially in size of input data for solving on a classical computer can be solved much faster, in polynomial time, if one uses a quantum computer [50]."

"If the Boolean algebra of operations is used, then the algorithm remains classical, regardless of what bits are used in this case, classical or quantum."

"The operations performed on the device are not reversible in sense of quantum mechanics, and time and resources required to execute the algorithm are approximately have the same order for quantum and classical computers."

"It is possible to obtain superiority in computational efficiency over classical algorithms if, instead of classical Boolean operations, quantum computing is applied."

The Quantum Machine Learning Models

"Artificial quantum neural network has been proposed in [51]."

"The weights are complex numbers (which change as the network is trained), so that each input quantum state is not only weighted in amplitude, but also phase shifted."

"Excluding nonlinear effects, quantum neural networks are identical to optical neural networks [52]."

"Optical quantum neural networks have a progress [53, 54]."

"In the known models of the quantum neural networks, the procedure for setting up a neural network consists of a sequence of classical operations of measuring and preparing states."

"Convolutional neural network (CNN) is a success case of quantum neural networks that has demonstrated its quantum advantage in recognition of complex objects (e.g [55, 56].)."

"As in a classical CNN, the QCNN hyperparameters such as the number of convolution and subsampling layers are fixed, and the unitaries themselves are learned [57]."

Platforms and Software Frameworks Supporting the QML

"The quantum devices D-Wave [58] and IBM Q [59] are the most developed quantum platforms that can support the QML: D-Wave has the status of an"analog quantum computer", as it is able to solve only a narrow range of the quantum annealing tasks, but, at the same time, they declare its capacity about 2000 qubits; IBM Q is a general purpose quantum computer on which arbitrary quantum algorithms can be run."

"The most promising kit is Qiskit, which allows us to manage the resources and adapt the developed ML applications for the specific quantum computers."

"Tensorflow Quantum library was selected for its main advantages: Flexibility, ready-to-use ML models, box application packages, scalability in hardware and software, large online community, and compatibility with Keras library."

Synthesis of a Stream Dataset for the Quantum Classifiers

"Existing network datasets, such as IEEEDataPort's IoT Network Intrusion Dataset [60], Stratosphere Lab's Malware on IoT Dataset [61] and Canberra's BoT-IoT Dataset [62], contain the network packets of two classes:" with attack" and"without attack"."

"The fields in the stream are the transformed fields of the packets entering that stream."

"In IoT Network Intrusion Dataset, packet fields like tcp.srcport, tcp.dstport, udp.srcport, udp.dstport, tcp.checksum.status, udp.checksum.status are merged in to srcport, dstport, ip.checksum.status to get rid of the Layer 4 protocol dependency."

"The tcp.stream and udp.stream fields are combined in to the stream field, which is used to group packets to the streams."

"An attack stream "Flooding" with equal intervals between the packets will give a low average deviation in the size of the intervals."

Data Coding for the QML

"Using the Cirq library, a qubit is created, which is placed in the circuit: Next, the IP address is converted into the number, on the basis of which the angle of rotation is set up, then the Pauli gate is added to the circuit."

"The rotation angle is the value into which the IP address is converted: The algorithm then starts working with the fields that contain numeric values."

"The value obtained in the previous step is utilized as a rotation angle."

The Experimental Study

"With the QML-based intrusion detection, the test bench has been constructed: Ubuntu operating system provides interoperability between the hardware and software; Python 3.7 programming language sets links the program and the operating system; Cirq library executes the quantum circuits; NVidia Cuda provides the faster emulation for quantum circuits; Tensorflow provides the platform for building ML method; Tensorflow Quantum contains the framework structures for the QML such as qubits, gates, schematics, and measurement operators."

"When using the conventional SVM, only HTTP Flooding and Port Scanning attacks are detected with high accuracy."

"QSVM and QCNN can be trained approximately twice faster on a big input, and this priority keeps growing if the input data volume enlarges more."

"The results of the experiments have demonstrated abilities of the QML against the conventional ML-based detectors when classifying the big volumes of input data."

Conclusion

"As the result of our research, the possibility of using the QML methods to solve the problem of analyzing big volumes of input data was considered."

"When we have a large-scale network with a big volume of security-relevant data, the QML-based intrusion detection makes the protection more efficient than a traditional ML approach."

"Comparison of the QML detectors built on the QSVM and QCNN classifiers against the conventional SVM and QCNN detectors has shown the promise of the quantum apparatus on big data inputs."

"The QML-based methods have surpassed the ML-based implementations both in accuracy and performance."

"Comparison of the conventional ML classifiers and the QML classifiers on huge stream datasets has shown significant superiority of the quantum approach (e.g., QSVM and QCNN classification accuracy is 98%)."

Analysis of a Huge Amount of Network Traffic Based on Quantum Machine Learning [63]

This is a machine-generated summary of:

Kalinin, M. O.; Krundyshev, V. M.: Analysis of a Huge Amount of Network Traffic Based on Quantum Machine Learning [63].

Published in: Automatic Control and Computer Sciences (2021).

Link to original: https://doi.org/10.3103/s014641162108040x

If you want to cite the papers, please refer to the original.

For technical reasons we could not place the page where the original quote is coming from.

Abstract-Summary

"A method for analyzing network traffic based on quantum machine learning is presented."

Quantum Machine Learning

"The basic idea of quantum machine learning is to merge machine learning and quantum computing techniques."

"Quantum machine learning uses hybrid methods involving both classical and quantum processing, where computationally complex routines are outsourced to a quantum device [64, 65]."

"Quantum computers DWave and IBM Q are the most elaborated."

"IBM Q is a project for the development of universal quantum computers that can execute arbitrary quantum algorithms."

"The instantaneous processing of huge amounts of data and the solution of optimization problems make quantum technologies one of the most promising tools in the field of artificial intelligence and machine learning."

"Researchers consider quantum machine learning to be one of the most promising areas in the field of quantum computing [66]."

"When processing large amounts of data, the use of quantum computing can achieve quadratic and even exponential acceleration compared to their classical counterparts."

Stream Dataset Generation

"The existing datasets, such as IEEEDataPort IoT Network Intrusion Dataset [67], Stratosphere Lab Malware on IoT Dataset [68], and NSW Canberra The BoT-IoT Dataset [69], contain network packets of two categories: "with attack" and "without attack."".

"We propose transforming the IoT Network Intrusion Dataset that contains six types of attacks and normal traffic into a stream dataset."

"The 'tcp.stream' and 'udp.stream' fields have been merged into the 'stream' field that is used to group packets into streams."

"The fields in the stream represent the transformed fields of the packets included in that stream."

"The intervals between packets in the stream are analyzed separately, which allows one to judge the frequency of packet sending."

"A flooding attack stream with equal intervals between packets will give a low average deviation of interval values."

The Coding Method Developed

"GridQubit(1,1) Next, the IP address is transformed into a number that is the basis for setting the rotation angle, and then a Pauli gate is added to the circuit."

"Circuit() >>> a =make_num_from_ip(i[0]) >>> a =make_angle(a) >>> C.append(cirq.rx(a)(qubit)) Then the algorithm starts working with the fields that contain numerical values."

"The rotation angle is generated for each of them, and the next Pauli gate is added to the circuit."

">>> for j in range(1,58): >>> angle=make_angle(i[j]) >>> C.append(cirq.rx(angle)(qubit)) Therefore, each stream field is transformed into a number from 0 to π, after which a qubit is created for each stream."

"The value obtained at the previous step is used as the rotation angle, and, then, the resulting quantum circuit is added to the list, which serves to train the classifier."

Experimental Results

"Nvidia CUDA provides faster emulation for running quantum circuits."

"Tensorflow Quantum contains basic structures such as qubits, logic elements, circuits, and measurement operators."

"This method is effective when dealing with big data, has no tendency to over-training, provides high accuracy when dealing with a large attribute space, and allows using the kernel trick [69]."

"When the classical SVM is used, only HTTP flooding and port scanning attacks are detected with high accuracy."

"When using the quantum SVM, the binary classification problem was solved with a high accuracy of 98%."

Conclusions

"The possibility of applying quantum-learning methods to solve the problem of Huge Data analysis is considered."

"The method for encoding a bit representation of network streams into a qubit representation is developed for quantum information processing."

"The experimental results show the superiority of quantum machine learning over classical machine learning in solving the problem of classification of a huge amount of network traffic."

"The use of quantum machine learning reduce the learning time more than twice."

[Section 1]

"Since 1969, when the first computer network was created by the U.S. Defense Advanced Research Projects Agency, a steady increase in the amount of network traffic has been observed."

"The emergence of Internet of Things devices, digitalization of production, implementation of next generation 5G/5G+ wireless networks, and development of streaming services have resulted in extremely high amounts of traffic recorded by network equipment vendors."

"Traditional intrusion detection systems (IDSs) based on signature analysis have been used successfully for a long time as protection against network attacks."

"The method for the incoming traffic mapping to attack patterns has several disadvantages when analyzing big data."

"When processing not just big data, but huge data $>10^6$, these systems display a decrease in the accuracy of network traffic classification, and an increase in the processing time of incoming packets."

"It is proposed to use quantum machine learning methods to analyze huge amounts of network traffic."

Hierarchical System Mapping for Large-Scale Fault-Tolerant Quantum Computing [70]

This is a machine-generated summary of:

Hwang, Yongsoo; Choi, Byung-Soo: Hierarchical system mapping for large-scale fault-tolerant quantum computing [70].

Published in: Quantum Information Processing (2021).

Link to original: https://doi.org/10.1007/s11128-021-03151-8

If you want to cite the papers, please refer to the original.

For technical reasons we could not place the page where the original quote is coming from.

Abstract-Summary

"It is practically infeasible to deal with such problems with conventional methods based on a non-structured description about quantum algorithm."

"To overcome the problems, we propose a fast method by using a hierarchically structured description about quantum algorithm which is much more compact than the conventional method."

"During the process, the dedicated computing regions and their interconnection are dynamically mapped onto a structured quantum computing system architecture."

"Since the combination of structured code and architecture provides a high degree of locality, it requires less SWAP chains, and hence, it does not increase the quantum computation depth more than expected."

Introduction and Background

"A non-modular QASM describes a quantum algorithm at the individual gate level."

"Most quantum computing system mappings have been performed by taking the non-modular QASM [71–77]."

"When, however, we turn our attention to a large-sized quantum algorithm, several practical problems arise with the non-modular QASM."

"As the size of a quantum algorithm increases, so does the size of the non-modular QASM."

"Due to the lack of classical storage and memory, we could not even attempt to generate the non-modular QASM for the more large-sized quantum algorithm."

"Even if we generate a non-modular QASM for a large-sized quantum algorithm successfully, it does not seem practical to perform the system mapping by reading such huge size QASM."

"With the modular QASM approach, it will be possible to generate and manage a QASM practically for a large-sized quantum algorithm."

Hierarchical System Mapping

"At a module, qubits are manipulated by the quantum instructions described in a modular QASM, and transmitted between modules through the communication bus."

"For the mapping of a module, we first need to allocate a physical space for the module next to the previously allocated module and arrange the qubits of QASM on that space as much as required."

"In a modular QASM, there are three kinds of quantum instructions: 1- and 2-qubit gate and module."

"Depending on quantum algorithms and their implementation, some terminal modules may take only one-parameter qubit."

"This decomposition can be defined as a one-parameter-qubit module in the quantum algorithm."

"Please note that for a multi-parameter-qubit terminal module, we have to allocate separate physical space and perform the mapping of the module on that place to keep the heart of the proposed mapping."

Discussion

"In terms of the complexity for finding the shortest path for qubit move to perform an arbitrary CNOT gate, that of the modular mapping is much smaller than that of the non-modular mapping."

"For a CNOT gate, in the non-modular mapping, the whole qubit layout is the target space for searching the shortest path, but in the modular mapping, the current module is the target qubit layout, not the whole system."

"The pre-analysis of the modular QASM, which may be required for optimizing the qubit mapping, can be done very efficiently than the non-modular QASM."

"We just mentioned that more qubit movements are required for the proposed modular QASM mapping."

"We need to say that for a CNOT gate over arbitrary two qubits, the modular mapping requires fewer qubit movements."

"The number of SWAP gates is increased in the non-modular mapping based on the monolithic qubit layout."

Summary of Performance Results

"We summarize the performance of the proposed modular mapping discussed in the previous section."

"It then compares the performance of the quantum computing models based on non-modular mapping and modular mapping in the small input size regime."

"Please note that non-modular mapping is applicable to this regime."

Conclusion

"We have proposed an efficient circuit mapping methodology for a large-scale quantum computing."

"Several times, it is practically infeasible to deal with quantum computing in the large-scale regime with the conventional method based on a non-modular QASM."

"Our future work is to apply the most realistic quantum computer architecture [78] and to improve the practicality of the mapping algorithm."

"We assume the bus works on demand and qubit passings are done on time."

Formalization of Metatheory of the Quipper Quantum Programming Language in a Linear Logic [79]

This is a machine-generated summary of:

Mahmoud, Mohamed Yousri; Felty, Amy P.: Formalization of Metatheory of the Quipper Quantum Programming Language in a Linear Logic [79].

Published in: Journal of Automated Reasoning (2019).

Link to original: https://doi.org/10.1007/s10817-019-09527-x

If you want to cite the papers, please refer to the original.

For technical reasons we could not place the page where the original quote is coming from.

Abstract-Summary

"We develop a linear logical framework within the Hybrid system and use it to reason about the type system of a quantum lambda calculus."

"We extend the system with a linear specification logic (SL) in order to reason about the linear type system of Quipper."

"We formalize the semantics of Proto-Quipper by encoding the typing and evaluation rules in the SL, and prove type soundness."

Introduction

"In the case of Hybrid, an intermediate level is introduced by inductively defining a specification logic (SL) in Coq, and OL judgments are encoded in the SL."

"Hybrid is presented in full detail in [80], including a minimal intuitionistic logic as an SL, and one of the first case studies proving subject reduction for big-step semantics of a fragment of a pure functional language known as Mini-ML."

"We discuss the same Proto-Quipper case study in that paper with a reference to the full Coq code, but the presentation covers only a small part, including the encoding of Proto-Quipper terms, the encoding of a few typing and reduction rules, and an abbreviated statement of the subject reduction theorem."

"One of the strengths of Hybrid as implemented in Coq is that it is straightforward to combine such direct encodings of OL syntax with encodings that use the HOAS and SL facilities provided by Hybrid."

Proto-Quipper

"The types and terms of Proto-Quipper are defined by the following grammars: Proto-Quipper distinguishes between quantum data types (T, U) and types (A, B) where the former is a subset of the latter, and similarly between quantum data terms (t) and terms (a, b, c)."

"Most types and term constructs come directly from the quantum lambda calculus, e.g., [81]."

"A circuit constant C represents a low-level quantum circuit, and a term (t, C, a) represents a circuit as Proto-Quipper data, where t is a structure representing a finite set of inputs to C, and similarly a represents a finite set of outputs."

"Q is a quantum context containing a finite set of quantum variables, typically the free quantum variables in a. Also a is a term and A is a type."

"The subtyping rule in [82] is restricted only to term variables."

Proto-Quipper Types

"We have proved that (valid A) implies (validT A), confirming that class T, U is a subclass of A, B. The last step in the formalization of the Proto-Quipper types is the development of the subtyping relation."

"This presentation of the subtyping rules for the bang operator make the formal proof of the transitivity of the subtyping relation much easier; in particular it avoids a lot of unneeded induction cases."

"The following specification ensures that whenever the top-level type constructor of the super-type is the bang operator, then the subtype should be too."

"We provide two important properties of the subtyping relation: reflexivity and transitivity: Note that reflexivity is subject to the validity of A, i.e., it belongs to the Proto-Quipper types."

"This concludes the Proto-Quipper types formalization, where we considered the formal development of valid Proto-Quipper types and the subtyping relation on them."

Two-Level Hybrid

"The first step towards the formalization of the linear specification logic in Coq is defining an inductive type oo for formulas given by the G and P grammars: where the atom constructor accepts an atomic formula of type atm and casts it into the formula type oo."

"This step is done using an inductive predicate definition as follows: where we use the variables IL to represent the intuitionistic context, and LL for the linear one."

"The predicate splitseq is used twice; once for the intuitionistic subgoals iL under the empty linear context, and once for the linear subgoals lL. When we discuss the rules for prog later, we will say that we must intuitionistically prove the goals in iL and linearly prove the goals in lL. The predicates seq and splitseq are defined using Coq's mutual induction."

Encoding Proto-Quipper Programs and Semantics in Hybrid

"We have two function abstractions, e.g., fun x => fun y => App y x. Since the lambda is only defined for functions of type exp -> exp (not exp -> exp -> exp), we have to make two applications of lambda in the way presented in the above definition in order to satisfy the typing condition of lambda."

"It is important to clarify that the formalization presented in the previous section does not guarantee that all instances of the type qexp are valid in Proto-Quipper. (It just defines expressions that we are interested in.) This will be done as part of the program context prog with the help of the constructor is_qexp."

"We start by presenting examples of syntax rules that define valid expressions inside Proto-Quipper: Constants of Proto-Quipper are unconditionally valid; the list of linear and intuitionistic subgoals are empty."

Type Soundness

"We formally verify the type soundness of Proto-Quipper by addressing three important properties: a type soundness under subtyping rule, inversion lemmas for Proto-Quipper values, and the subject reduction theorem."

"Similar results have been proved for other Proto-Quipper expressions, e.g., functions, function applications, circuit conversion constants and so on. Reduction rules, i.e., operational semantics, are crucial for any programming language, defining how valid expressions in a language are simplified until a non-reducible expression is reached, i.e. a value."

"It is proved by the height of the proof of the sequent containing the typing judgment, followed by a case analysis of 17 reduction rules, each of which requires at least 10 major subgoals due to case analysis and inversions."

Adequacy

"Adequacy of syntax encoding, also called representational adequacy, is discussed for the lambda calculus as an OL in Hybrid in [83] and proved in detail in [84], while adequacy for a fragment of a functional programming language known as Mini-ML is proved in [80]."

"Proto-Quipper contains the lambda calculus as a sublanguage, and representational adequacy for the full language is a straightforward extension of these other results."

"To adequately represent OL syntax, we must rule out exotic functions, i.e., functions that do not encode OL lambda terms."

"Proving representational adequacy requires defining an encoding function between OL terms and their representation in Hybrid, and showing that this function is a bijection."

"Proving the adequacy of the encoding of inference rules is similar."

"For all judgments except well-formedness, internal adequacy lemmas must be proven."

"They are called internal adequacy lemmas because they can be formalized in Hybrid and they are an important part of the general adequacy proofs for these judgments."

"The following theorem expresses internal adequacy for the Proto-Quipper typing judgment."

Conclusion

"This work involved encoding a linear specification logic and carrying out a large case study in Hybrid, in the sense that we encode and reason about the complete Proto-Quipper specification, the most complex OL considered so far, and we prove type soundness, one of the central results in [82]."

"The ordered linear logic (OLF) mentioned earlier is one such example, where type preservation for a continuation-based abstract machine for the functional programming language Mini-ML (an OL that is simpler than Proto-Quipper) is proven [80] following the statement in [85]."

"Of OLs that benefit from a framework based on linear logic are those with imperative features."

"Examples involving mutable state have motivated a variety of proposals for frameworks based on linear logics that support HOAS."

"Type soundness proofs for various OLs has been a common benchmark for logical frameworks supporting HOAS and implementing an intuitionistic logic, starting with some of the earliest logical frameworks like LF [86]."

References

1. Weder B, Barzen J, Beisel M, Leymann F (2023) Provenance-preserving analysis and rewrite of quantum workflows for hybrid quantum algorithms. SN Computer Science. https://doi.org/10.1007/s42979-022-01625-9
2. Eder J, Liebhart W (1996) Workflow Recovery. In: Proceedings of the international conference on cooperative information systems. IEEE, pp 124–134
3. Leymann F (1995) Supporting business transactions via partial backward recovery in workflow management systems. In: Datenbanksysteme in Büro, Technik und Wissenschaft. Springer, pp 51–70
4. Leymann F, Barzen J (2021) Hybrid quantum applications need two orchestrations in superposition: a software architecture perspective. arXiv:2103.04320
5. McClean JR, Romero J, Babbush R, Aspuru-Guzik A (2016) The theory of variational hybrid quantum-classical algorithms. New J Phys 18:2
6. Brayton RK, Hachtel GD, Sangiovanni-Vincentelli AL (1990) Multilevel logic synthesis. Proc IEEE 78:264–300
7. Weder B, Barzen J, Leymann F (2021) MODULO: modeling, transformation, and deployment of quantum workflows. In: Proceedings of the 25th international enterprise distributed object computing workshop (EDOCW). IEEE, pp 341–344
8. Weder B, Breitenbücher U, Leymann F, Wild K. Integrating quantum computing into workflow modeling and execution. In: Proceedings of the 13th IEEE/ACM international conference on utility and cloud computing (UCC). IEEE; 2020, p. 279–291
9. Camunda: Camunda BPMN Modeler 2022. https://camunda.com/products/camunda-bpm/modeler
10. IBM: Qiskit runtime; 2022. https://github.com/Qiskit-Partners/qiskit-runtime
11. AWS: Amazon Braket Hybrid Jobs User Guide 2022. https://docs.aws.amazon.com/braket/latest/developerguide/braket-jobs.html
12. Khan SU, Awan AJ, Vall-Llosera G (2019) K-means clustering on noisy intermediate scale quantum computers. arXiv:1909.12183
13. Di Penta M, Esposito R, Villani ML, Codato R, Colombo M, Di Nitto E (2006) WS Binder: a framework to enable dynamic binding of composite web services. In: Proceedings of the 2006 international workshop on Service-oriented software engineering, pp 74–80
14. Reiff-Marganiec S, Yu HQ, Tilly M (2007) Service selection based on non-functional properties. In: International Conference on Service-Oriented Computing (ICSOC). Springer, pp 128–138
15. Song X, Dou W, Chen J (2011) A workflow framework for intelligent service composition. Futur Gener Comput Syst 27(5):627–636
16. Zimmermann M, Breitenbücher U, Képes K, Leymann F, Weder B (2020) Data flow dependent component placement of data processing cloud applications. In: Proceedings of the IEEE international conference on cloud engineering (IC2E). IEEE, pp 83–94
17. Bucchiarone A, Marconi A, Pistore M, Raik H. Dynamic adaptation of fragment-based and context-aware business processes. In: Proceedings of the 19th international conference on web services (ICWS). IEEE, 2012. p. 33–41
18. Mundbrod N, Grambow G, Kolb J, Reichert M (2015) Context-aware process injection: enhancing process flexibility by late extension of process instances. In: On the Move to Meaningful Internet Systems (OTM). Springer, pp 127–145
19. Képes K, Breitenbücher U, Sáez SG, Guth J, Leymann F, Wieland M (2016) Situation-aware execution and dynamic adaptation of traditional workflow models. In: Proceedings of the 5th European conference on service-oriented and cloud computing (ESOCC). Springer, pp 69–83
20. Agrawal R, Gunopulos D, Leymann F (1998) Mining process models from workow logs. In: International Conference on Extending Database Technology. Springer, Berlin, pp 467–483
21. Van der Aalst W (2012) Process mining. Commun ACM 55(8):76–83

22. Van Dongen BF, de Medeiros AKA, Verbeek H, Weijters A, van der Aalst W (2005) The ProM framework: a new era in process mining tool support. In: International Conference on Applications and Theory of Petri Nets. Springer, pp 444–454
23. Cohen-Boulakia S, Biton O, Cohen S, Davidson S (2008) Addressing the provenance challenge using ZOOM. Concurr Comput Pract Exp 20(5):497–506
24. Schumm D, Leymann F, Streule A (2010) Process views to support compliance management in business processes. In: International Conference on Electronic Commerce and Web Technologies. Springer, pp 131–142
25. Serrano MA, Sánchez LE, Santos-Olmo A, García-Rosado D, Blanco C, Barletta VS, Caivano D, Fernández-Medina E (2023) Minimizing incident response time in real-world scenarios using quantum computing. Softw Qual J. https://doi.org/10.1007/s11219-023-09632-6
26. Hariyanti E, Djunaidy A, Siahaan DO (2018) A conceptual model for information security risk considering business process perspective. In: 2018 4th International Conference on Science and Technology (ICST). IEEE, Yogyakarta, pp 1–6. https://doi.org/10.1109/ICSTC.2018.8528678
27. Szwaczyk S, Wrona K, Amanowicz M (2018) Applicability of risk analysis methods to risk-aware routing in software-defined networks. In: 2018 International Conference on Military Communications and Information Systems (ICMCIS). IEEE, Warsaw, pp 1–7. https://doi.org/10.1109/ICMCIS.2018.8398688
28. Ruan K (2017) Introducing cybernomics: a unifying economic framework for measuring cyber risk. Comput Secur 65:77–89. https://doi.org/10.1016/j.cose.2016.10.009
29. Alshawabkeh M, Li X, Sullabi M (2019) New information security risk management framework as an integral part of project life cycle. In: Proceedings of the 2019 5th International Conference on Humanities and Social Science Research (ICHSSR 2019). Atlantis Press, Paris. https://doi.org/10.2991/ichssr-19.2019.24
30. Ross, R., Pillitteri, V., Graubart, R., Bodeau, D., & McQuaid, R. (2019). Developing cyber resilient systems: a systems security engineering approach. National Institute of Standards and Technology: Technical report
31. Turskis Z, Goranin N, Nurusheva A, Boranbayev S (2019) Information security risk assessment in critical infrastructure: a hybrid MCDM approach. Informatica 30(1):187–211. https://doi.org/10.15388/Informatica.2019.203
32. Bhardwaj A, Sapra V (2021) Security incidents & response against cyber attacks. Springer. https://doi.org/10.1007/978-3-030-69174-5
33. Clairambault P, DeVisme M, Winskel G (2019) Game semantics for quantum programming. Proc ACM Program Lang 3(POPL):1–29
34. Svore K, Roetteler M, Geller A, Troyer M, Azariah J, Granade C, Heim B, Kliuchnikov V, Mykhailova M, Paz A (2018) Q#. In: Proceedings of the Real World Domain Specific Languages Workshop 2018. RWDSL2018. ACM Press, New York, pp 1–10. https://doi.org/10.1145/3183895.3183901
35. Liu S, Wang X, Zhou L, Guan J, Li Y, He Y, Duan R, Ying M (2018) Qsi >: a quantum programming environment. In: Lecture Notes in Computer Science (including subseries Lecture Notes in Artificial Intelligence and Lecture Notes in Bioinformatics), vol 11180(LNCS). Springer, pp 133–164. https://doi.org/10.1007/978-3-030-01461-2_8
36. Steiger DS, Häner T, Troyer M (2018) ProjectQ: an open source software framework for quantum computing. Quantum 2:49. arXiv:1612.08091. https://doi.org/10.22331/q-2018-01-31-49
37. Abraham H et al (2019) Qiskit: an open-source framework for quantum computing. Zenodo. https://doi.org/10.5281/zenodo.2562111
38. Piattini M, Serrano M, Perez-Castillo R, Petersen G, Hevia JL (2021) Toward a quantum software engineering. IT Prof 23(1):62–66. https://doi.org/10.1109/MITP.2020.3019522
39. Ahmad A, Maynard SB, Shanks G (2015) A case analysis of information systems and security incident responses. Int J Inf Manag 35(6):717–723. https://doi.org/10.1016/j.ijinfomgt.2015.08.001
40. Azad U, Sinha A (2023) qLEET: visualizing loss landscapes, expressibility, entangling power and training trajectories for parameterized quantum circuits. Quantum Inf Process. https://doi.org/10.1007/s11128-023-03998-z

41. Romero J, Babbush R, McClean JR, Hempel C, Love PJ, Aspuru-Guzik A (2018) Strategies for quantum computing molecular energies using the unitary coupled cluster ansatz. Quantum Sci Technol 4:014008. https://doi.org/10.1088/2058-9565/aad3e4
42. Kandala A, Mezzacapo A, Temme K, Takita M, Brink M, Chow JM, Gambetta JM (2017) Hardware-efficient variational quantum eigensolver for small molecules and quantum magnets. Nature 549:242–246. https://doi.org/10.1038/nature23879
43. Tang HL, Shkolnikov V, Barron GS, Grimsley HR, Mayhall NJ, Barnes E, Economou SE (2021) Qubit-ADAPT-VQE: an adaptive algorithm for constructing hardware-efficient ansätze on a quantum processor. PRX Quantum 2:020310. https://doi.org/10.1103/PRXQuantum.2.020310
44. Meuli G, Soeken M, De Micheli G (2022) Xor-And-inverter graphs for quantum compilation. npj Quantum Inform 8. https://doi.org/10.1038/s41534-021-00514-y
45. Langenberg B, Pham H, Steinwandt R (2020) Reducing the cost of implementing the advanced encryption standard as a quantum circuit. IEEE Trans Quantum Eng 1:1–12
46. Amy M, Maslov D, Mosca M, Roetteler M (2013) A meet-in-the-middle algorithm for fast synthesis of depth-optimal quantum circuits. IEEE Trans CAD Integ Circuit Syst 32:818–830
47. Kalinin M, Krundyshev V (2022) Security intrusion detection using quantum machine learning techniques. J Comput Virol Hacking Techniques 19:125–136. https://doi.org/10.1007/s11416-022-00435-0
48. Lavrova D, Poltavtseva M, Shtyrkina A (2018) Security analysis of cyber-physical systems network infrastructure. In: Proceedings of 2018 IEEE Industrial Cyber-Physical Systems, pp 818–823
49. Pan S, Morris T, Adhikari U (2015) Developing a hybrid intrusion detection system using data mining for power systems. IEEE Trans Smart Grid 6:3104–3113
50. Simon DR (1997) On the power of quantum computation. SIAM J Comput 26(5):1474–1483
51. Kak S (1995) On quantum neural computing. Inf Sci 83(3–4):143–160
52. Shariv I, Friesem AA (1989) All-optical neural network with inhibitory neurons. Opt Lett 14(10):485
53. Inagaki T, Inaba K, Hamerly R, Inoue K, Yamamoto Y, Takesue H (2016) Large-scale ising spin network based on degenerate optical parametric oscillators. Nat Photonics 10(6):415–419
54. McMahon PL, Marandi A, Haribara Y, Hamerly R, Langrock C, Tamate S, Inagaki T, Takesue H, Utsunomiya S, Aihara K, Byer RL, Fejer MM, Mabuchi H, Yamamoto Y (2016) A fully programmable 100-spin coherent ising machine with all-to-all connections. Science 354(6312):614–617
55. LeCun Y, Bengio Y (2003) Convolutional networks for images, speech, and time series, pp 276–279
56. Krizhevsky A, Sutskever I, Hinton GE (2017) Imagenet classification with deep convolutional neural networks. Commun ACM 60(6):84–90
57. Cong I, Choi S, Lukin MD (2019) Quantum convolutional neural networks. Nat Phys 15(12):1273–1278
58. Ding Y, Lamata L, Martín-Guerrero J, Lizaso MSE, Chen X, Orus R, Solano EMS (2019) Towards prediction of financial crashes with a D-wave quantum computer. arXiv:1904.05808
59. Santos AC (2016) The IBM quantum computer and the IBM quantum experience. arXiv:1610.06980
60. Kang H, Ahn DH, Lee GM, Yoo JD, Park KH,Kim HK (2019) IoT network intrusion dataset. http://ieee-dataport.org/open-access/iot-network-intrusion-dataset
61. Garcia S, Parmisano A, Erquiaga MJ (2020) IoT-23: A labeled dataset with malicious and benign IoT network traffic (Version 1.0.0). www.stratosphereips.org/datasets-iot23
62. Koroniotis N, Moustafa N, Sitnikova E, Turnbull B (2018) Towards the development of realistic botnet dataset in the Internet of Things for network forensic analytics: Bot-IoT dataset. arXiv:1811.00701
63. Kalinin MO, Krundyshev VM (2021) Analysis of a huge amount of network traffic based on quantum machine learning. Autom Control Comput Sci 55:1165–1174. https://doi.org/10.3103/s014641162108040x

64. Wiebe N, Braun D, Lloyd S (2012) Quantum algorithm for data fitting. Phys Rev Lett 109(5):050505
65. Bisio A, D'Ariano GM, Perinotti P, Sedlák M (2011) Quantum learning algorithms for quantum measurements. Phys Lett A 375(39):3425–3434. https://doi.org/10.1016/j.physleta.2011.08.002
66. Biamonte J, Wittek P, Pancotti N, Rebentrost P, Wiebe N, Lloyd S (2017) Quantum machine learning. Nature 549(7671):195–202
67. Kang H, Ahn DH, Lee GM, Yoo JD, Park KH, Kim HK (2019) IoT Network Intrusion Dataset. IEEE Dataport. https://doi.org/10.21227/q70p-q449
68. Parmisano S, Garcia M, Erquiaga M (2020) Stratosphere laboratory. A labeled dataset with malicious and benign IoT network traffic. January 22th. https://www.stratosphereips.org/datasets-iot23. Cited December 25, 2020
69. Koroniotis N, Moustafa N, Sitnikova E, Turnbull B (2019) Towards the development of realistic botnet dataset in the Internet of Things for network forensic analytics: Bot-IoT dataset. Future Gen Comput Syst 100:779–796. https://doi.org/10.1016/j.future.2019.05.041
70. Hwang Y, Choi B-S (2021) Hierarchical system mapping for large-scale fault-tolerant quantum computing. Quantum Inf Process 20. https://doi.org/10.1007/s11128-021-03151-8
71. Saeedi M, Wille R, Drechsler R (2012) Synthesis of quantum circuits for linear nearest neighbor architectures, pp 1–14
72. Pedram M, Shafaei A (2016) Layout optimization for quantum circuits with linear nearest neighbor architectures. IEEE Circ Syst Mag 16(2):62–74
73. Zulehner A, Paler A, Wille R (2019) An efficient methodology for mapping quantum circuits to the IBM QX architectures. IEEE Trans Comput Aided Des Integr Circ Syst 38(7):1226–1236
74. Li G, Ding Y, Xie Y (2019) Tackling the Qubit mapping problem for NISQ-era quantum devices. In: The twenty-fourth international conference. ACM Press, New York, pp 1001–1014
75. Guerreschi GG, Park J (2018) Two-step approach to scheduling quantum circuits. Quant Sci Technol 3(4):045003
76. Murali P, Baker JM, Abhari AJ, Chong FT, Martonosi M (2019) Noise-adaptive compiler mappings for noisy intermediate-scale quantum computers. In: The twenty-fourth international conference. ACM Press, New York, pp 1015–1029
77. Lao, L., van Someren, H., Ashraf, I., Almudever, C.G.: Timing and resource-aware mapping of quantum circuits to superconducting processors (2019). https://arxiv.org/abs/1908.04226
78. Lin CC, Sur-Kolay S, Jha NK (2015) PAQCS: Physical design-aware fault-tolerant quantum circuit synthesis. IEEE Trans Very Large Scale Integr (VLSI) Syst 23(7):1221–1234
79. Mahmoud MY, Felty AP (2019) Formalization of metatheory of the Quipper quantum programming language in a linear logic. J Autom Reason 63:967–1002. https://doi.org/10.1007/s10817-019-09527-x
80. Felty AP, Momigliano A (2012) Hybrid: A definitional two-level approach to reasoning with higher-order abstract syntax. J Autom Reason 48(1):43–105
81. Selinger P, Valiron B (2006) A lambda calculus for quantum computation with classial control. Math Struct Comput Sci 16(3):527–552
82. Ross NJ (2015) Algebraic and logical methods in quantum computation. PhD thesis, Dalhousie University, August (2015). arXiv:1510.02198 [quant-ph]
83. Ambler S, Crole RL, Momigliano A (2002) Combining higher order abstract syntax with tactical theorem proving and (co)induction. In: 15th International Conference on Theorem Proving in Higher-Order Logics (TPHOLs), Lecture Notes in Computer Science. Springer, pp 13–30
84. Crole RL (2011) The representational adequacy of hybrid. Math Struct Comput Sci 21(3):585–646
85. Cervesato I, Pfenning F (2002) A linear logical framework. Inf Comput 179(1):19–75
86. The Twelf Project (2009) Introduction to Twelf: proving metatheorems about the STLC. http://twelf.org/wiki/Proving_metatheorems:Proving_metatheorems_about_the_STLC. Accessed 1 Oct 2016

Chapter 3
Post-Quantum Cryptography

Introduction by the Editor

This chapter deals with post-quantum cryptography (PQC) whose algorithms and protocols have been designed and implemented to protect sensitive and confidential data against attacks or cracks with quantum computers. Using Shor's algorithm on a large-enough quantum computer or high-performance cloud, a malicious attacker could cryptanalyze any Public Key Cryptosystems like RSA/ECC public key and generate fake or duplicate digital signatures in a millisecond. NIST is making a lot of efforts to standardize PQC algorithms through competitions. Further, Quantum Key Distribution (QKD) using the BB84 protocol has been designed and developed for applications. Let us take up a new topic like proxy signature scheme which has been explored in this chapter. A quantum secure ID-based cryptographic encryption based on group rings is proven secure to use over conventional crypto encryption algorithms. The existing world faces several digital information breaches due to weak utilization of confidentiality preserving techniques. In this regard, a novel digital contents privacy scheme based on the quantum harmonic oscillator and Schrodinger paradox has been studied.

In this chapter, there is a growing threat imposed on public key cryptography (PKC). The threat or risk comes from a fast growth on quantum computers which start cracking PKC. That's why we need post-quantum algorithms to protect sensitive data. Of course, it highlights the recent efforts of the National Institute of Standards and Technology (NIST) USA in standardizing post-quantum algorithms and discusses the challenges associated with integrating these algorithms or protocols into existing cryptographic libraries. We describe how the researchers integrated the lattice-based post-quantum algorithms into the Crypto++ cryptography library. Further, we discuss the challenges or open problems during the integration process in a crypto system, including those related to the mathematical complexity of lattice-based algorithms or models.

V. Natarajan (ed.), *Quantum Artificial Intelligence*,
https://doi.org/10.1007/978-981-96-5051-4_3

Although we state the importance of secure key management in cryptographic systems and introduce quantum cryptography, specifically the Quantum Key Distribution (QKD) protocol, as a method for securely transmitting and sharing secret keys between entities. By integrating QKD with conventional cryptography algorithms, the paper aims to enhance or improve the security of data transmission to a remarkably high level. Further. we introduce a novel digital signature scheme with proxy delegation and message recovery features in an identity-based context. Even though, this scheme or algorithm is designed and analyzed to address the need for secure ownership enforcement tools in real-world industry applications such as digital property transfer.

As the threats and risks of large-scale quantum computers capable of breaking conventional cryptographic protocols looms, the development of quantum-safe key exchange protocols (KEX) has become imperative. Next, this research conducts a systematic mapping study of post-quantum hybrid KEX, which amalgamate classical cryptographic key exchange protocols with novel post-quantum schemes.

Another interesting topic, Physical unclonable functions (PUFs) have emerged as a promising method for authenticating electronic devices. Extending this concept to quantum devices, several proposals have introduced classical readout quantum PUFs (CR-QPUFs), leveraging single-qubit gates to establish a secure fingerprint of quantum devices. In this study, we formalize the CR-QPUF concept within the statistical query (SQ) model and demonstrate its susceptibility to security breaches when adversaries possess SQ access to the CR-QPUF. Then we show that a malicious attacker could learn CR-QPUF characteristics by low-order polynomial regression techniques.

In this chapter, we cover up a topic of Quantum Refereed Games (ORG). It determines complexity theoretic aspects of two competing players who transfer quantum states to a referee who will measure the two states and evaluate the winner among the two players. Based on the player's strategy and high probability score of players, the decision will be taken on the game. It is shown that the complexity class QRG(1) contains QMA U co-QMA where QMA stands Quantum Merlin Arthur. In the subsection of Quantum information and Quantum circuits of QRG, we discuss the natural representation of quantum gates where the research studies explore several gates like Hadamard gate, Phase gate, Toffoli gate, Ancillary qubit gate and Erasure gate. This QRG model has practical applications such as electoral winners, bidding, and sports.

Next, we focus on a post-quantum fuzzy commitment (PQFC) protocol which is proven to be secure on a random oracle model. Then we conduct a comprehensive analysis to demonstrate the protocol's functionality and security features, including memoryless-effortless authentication mechanisms, user anonymity, risk analysis, mutual authentication, and resilience against various attacks such as biometric template tampering, stolen smart card, and privileged interior attacks. Subsequently, we estimate the protocol's storage, memory optimization, computation, communication, and storage costs, comparing them against existing crypto protocols. Due to the development of quantum computers, there is a threat or risk to classical Crypto algorithms such as RSA, Diffie-Hellman, and Elliptic Curve Diffie-Hellman for usage. The latest research explores quantum-based encryption methods. Then we outline the Authenticated Quantum-Secure Key Agreement and Message Encryption (AKAME) scheme, designed to offer optimal key sizes and decryption times. The scheme aims to establish a robust key agreement between two entities without the need for reconciliation mechanisms. Leveraging the Ring-Learning with Error (RLWE) problem, a lattice-based hard problem, AKAME ensures underlying security. The AKAME scheme can provide both 118 bits or 254 bits classical security and 102 bits or 241bits quantum security.

The healthcare system uses the Internet of Things (IoT) for the digital environment. Nowadays, PQFC plays an important role in healthcare. It protects biometric templates. In this book, we discuss an efficient method of biometric template protection using two factor authentication protocol for IoT-enabled healthcare ecosystems in pos-quantum computing environments. This kind of protocol provides security and privacy like resistance to tampering and theft of stored biometric templates, stolen smart cards and privileged interior attacks. During the performance of this protocol, we determine the metrics such as the storage requirements, communication costs, and computational complexities. The total computational time cost of the protocol is represented by a summation of multiplication modulo, vector addition modulo and one-way hash function.

In this chapter, classical algorithms (RSA, ECDSA) alongside hash-based signature schemes, specifically Winternitz one-time signature (W-OTS) and Merkle signature (MSS), we conduct a comprehensive review and analysis. Both W-OTS and MSS are hash based signatures. Our analysis focuses on their security strength and efficiency in terms of key generation time, signature generation, and verification time. The Lightweight Cryptography (LWC) competition was initiated by The National Institute of Standards and Technology (NIST) to standardize lightweight cryptographic algorithms. While submissions are required to be quantum-safe for long-term security, this aspect was not explicitly included in the submission requirements. Consequently, many candidates, such as sESTATE and Elephant, do not address security against quantum attacks. Both sESTATE and Elephant belong to LWC.

At the end of Chap. 3, we introduce an image encryption scheme that leverages the fundamental concepts from quantum harmonic oscillators and Schrödinger's paradox to provide heightened security. There are digital information breaches or attacks due to weak encryption algorithms. That's why quantum harmonic oscillators and Schrödinger's paradox are helpful to industry applications. Furthermore, we authenticate the effectiveness of our technique through comprehensive security performance analysis, comparing the results with existing benchmark schemes.

Machine Generated Summaries

Disclaimer: The summaries in this chapter were generated from Springer Nature publications using extractive AI auto-summarization: An extraction-based summarizer aims to identify the most important sentences of a text using an algorithm and uses those original sentences to create the auto-summary (unlike generative AI). As the constituted sentences are machine selected, they may not fully reflect the body of the work, so we strongly advise that the original content is read and cited. The auto generated summaries were curated by the editor to meet Springer Nature publication standards. To cite this content, please refer to the original papers.

Machine generated keywords: cryptography, postquantum, encryption, scheme, signature, security, digital, public key, digital signature, cryptographic, standardization, cryptographic algorithm, public, signature scheme, latticebase.

Implementing Post-Quantum Cryptography for Developers [1]

This is a machine-generated summary of:

Hekkala, Julius; Muurman, Mari; Halunen, Kimmo; Vallivaara, Visa: Implementing Post-quantum Cryptography for Developers [1].

Published in: SN Computer Science (2023).

Link to original: https://doi.org/10.1007/s42979-023-01724-1

If you want to cite the papers, please refer to the original.

For technical reasons we could not place the page where the original quote is coming from.

Abstract-Summary

"Post-quantum algorithms have been designed for the purpose of protecting sensitive data against attacks with quantum computers."

"When implementing complex cryptographic algorithms, developers commonly use cryptographic libraries in their solutions to avoid mistakes."

"Most of the open-source cryptography libraries do not yet have post-quantum algorithms integrated in them."

"We chose a C++ cryptography library, Crypto++, and created a fork where we integrated four lattice-based post-quantum algorithms."

"The performance of the integrated algorithms was overall good, but the integration process had its challenges, many of which were caused by the mathematical complexity of lattice-based algorithms."

"Different open-source implementations of post-quantum algorithms will be essential to their easier use for developers."

Introduction

"A modern computer is unable to solve these problems in practice but the development of quantum computers will make both of these public key cryptographic algorithms vulnerable."

"Post-quantum cryptographic algorithms are designed for this purpose."

"Post-quantum algorithms have rarely been implemented to open-source cryptographic libraries up to this point."

"We discuss the process of implementing post-quantum cryptographic algorithms into a programming language library."

"We chose Crypto++, which is a C++ language general use cryptographic library with a wide variety of different algorithms and created a fork of it to integrate post-quantum algorithms."

"To extend our work, in addition to improving the code in the fork we integrated another post-quantum algorithm into the fork, and used Known Answer Tests to further investigate the validity of the implementation."

"We dive into the background of post-quantum algorithms and open-source implementations, and then, we describe the work done."

Background

"Quantum cryptography means cryptographic algorithms that run on quantum computers, whereas post-quantum cryptography means algorithms that run on a classical computer but protect the secured data against attacks coming from quantum computers."

"The post-quantum cryptographic algorithms are designed to run on classical computers but meant to resist attacks from both classical and quantum computers."

"Different post-quantum algorithms are based on complex mathematical problems, such as lattices, coding theory or multivariate polynomials."

"Most of the largest open-source cryptography libraries have not yet implemented any post-quantum algorithms."

"One of the most used open-source cryptographic libraries, OpenSSL, has stated that they will not be adding any post-quantum algorithms to the library before the algorithms have been standardized [2]."

"It is important to study the implementation of post-quantum algorithms in advance to see how they actually work as part of a library in practice and what are the challenges in the implementation process."

Algorithm Integration

"The biggest benefit of the design in FrodoKEM is that in the future, some weaknesses could be found in the special cases of lattices that, e.g., Kyber and SABER are based on, and that these weaknesses would not be relevant in the case of generic lattices [3]. When integrating algorithms into a software library, there are certain things to be taken into account."

"The code structure of the new algorithms and the way they are integrated to the fork resemble the other algorithms in the library."

"We use library specific methods where applicable, and aim to produce little overhead in the code when integrating the algorithms to the fork."

"When making updates to the Dilithium implementation in the fork after having succesfully integrated the algorithm to the fork, the size of a vector was not correctly updated."

Results of Testing and Analysis

"We also studied the performance of the algorithms and compared the reference implementations to the performance with different compiler options."

"During their standardization process, NIST required each of the submission packages to include a file with Known Answer Test (KAT) values that can be used to determine the correctness of the algorithm implementations [4]."

"We implemented the algorithms PQCgenKAT_sign() for Dilithium and PQCgenKAT_kem() for Kyber, SABER and Frodo-KEM which were all modified from the reference implementations provided by the algorithm designers to the NIST standardization process."

"After integrating Frodo-KEM using SHAKE based on the implementation of the algorithm designers, we also tested the implementation using KAT values available on their GitHub repository."

"When at first compiling the implementations with default Crypto++ library options on Linux, especially the performance of SABER was a lot worse, and that is also prevalent in the table, the runtime basically doubling at times."

Discussion

"It is essential that post-quantum algorithms are implemented to open-source libraries, so that the threshold for developers and organizations to migrate to post-quantum or hybrid algorithms will lower."

"Implementing post-quantum, in our case lattice-based algorithms, is not simple."

"Without the reference implementations, integrating this many algorithms into our fork would not have been possible."

"It can be argued, that starting to use the post-quantum algorithms is important, so that they can be integrated into different systems and any overlooked mistakes are found as early as possible."

"The performance of the integrated algorithms in the fork was in the similar area when comparing to the original reference implementations—meaning that the integration into the library did not add much overhead into the implementation."

"When we chose the algorithms to be implemented, the third round of the NIST post-quantum standardization process was still ongoing and none of the algorithms had been standardized."

Conclusions

"In our paper, we have demonstrated how to integrate some of the candidate algorithms from the NIST post-quantum cryptography standardization contest to a widely used programming library."

"In our work one of the main goals was to make the new algorithms as intuitive and easy to use as possible for the developers utilising the Crypto++ library."

"The complexity of the cryptographic algorithms (especially PQC algorithms) adds to this and it is still necessary to improve on the usability of the libraries as well as the understanding of developers on the possibilities and limitations of the implementations."

"Open-source cryptographic libraries will play an important role in enabling developers to use post-quantum cryptography."

"Further research on the security of the algorithms as well as new implementations in different environments are necessary so that the confidence in the algorithms and usage of post-quantum algorithms rises."

Challenges of Post-Quantum Digital Signing in Real-World Applications: A Survey [5]

This is a machine-generated summary of:

Tan, Teik Guan; Szalachowski, Pawel; Zhou, Jianying: Challenges of post-quantum digital signing in real-world applications: a survey [5].

Published in: International Journal of Information Security (2022).

Link to original: https://doi.org/10.1007/s10207-022-00587-6

If you want to cite the papers, please refer to the original.

For technical reasons we could not place the page where the original quote is coming from.

Abstract-Summary

"Using Shor's algorithm [6, 7] on a large-enough quantum computer, an attacker can cryptanalyze any RSA/ECC public key and generate fake digital signatures in seconds."

"Besides understanding the digital signing usage, we compare the applications' signing requirements against all six NIST's Post-Quantum Cryptography Standardization round 3 candidate algorithms."

"This is done through a proposed framework where we map out the suitability of each algorithm against the applications' requirements in a feasibility matrix."

"Using the matrix, we identify improvements needed for all 14 applications to have a feasible post-quantum secure replacement digital signing algorithm."

Introduction

"NIST is spearheading the efforts through a Post-Quantum Cryptography (PQC) Standardization competition [8] to solicit and evaluate possible post-quantum secure digital signature algorithms for use beyond 2030."

"Our approach to investigate how digital signatures can continue to be used in the post-quantum era is a three-step process to (i) identify applications that use digital signatures; (ii) collate the digital signing operating requirements and constraints of these applications; and (iii) cross-reference the requirements against NIST's PQC Standardization round 3 algorithms."

"While we recognize that there are other techniques in achieving post-quantum digital signing [9–11], we choose to only use all six candidate algorithms from the NIST PQC Standardization round 3 for our evaluation matrix as all six have been thoroughly studied and at least one of them will be made into a standard and be widely adopted."

“We do a systematic literature review on Google Scholar of research related to application testing and implementations with post-quantum digital signatures and exclude algorithms that are not part of NIST’s round 3 candidates.”

Digital Signing in Practice

“While the majority of the cryptographic algorithms used in EMV are symmetric key-based, EMV relies on asymmetric key digital signatures in the static data authentication (SDA) and dynamic data authentication (DDA) protocols to reduce payment card fraud.”

“Impact Both EMV-SDA and EMV-DDA rely on digital signing for authentication of the card by the merchant terminal.”

“Impact The CA relies on digital signing to ensure the integrity of the identity and public key in the certificate, and this trust is inferred for any application relying on the certificates within the chain for authentication or non-repudiation.”

“A compromise of the digital signature scheme would be worse as attackers can create fake certificates and revocations that violate the trust and cause complete failure to any authentication or non-repudiation processes, without any means of remediation through key re-issuance.”

“Impact PDF signing relies on digital signing to provide document integrity (for PDF-AES) and signer non-repudiation (for PDF-QES).”

Digital Signing Requirements Framework

“If a signature scheme determines that key sizes need to be higher than 10,000 bits in size to achieve 128-bit security, but the application can only support key sizes of 2048 bits due to restrictions within the platform, then the signature scheme is not suitable for the application, regardless of the security or operational efficiency.”

“Applications that require key and signature sizes to be smaller than 2 Kbits impose the highest constraints.”

“Applications that can support such key or signature sizes are considered to impose a small constraint.”

“Applications that require to perform signing functions on these platforms impose the highest constraints, while the signature schemes that can run on these platforms are those that provide the most flexibility.”

“We classify applications that require to run on these platforms as having small constraints and signature schemes that can run on these platforms to be moderately flexible.”

Applying the Framework

"We attempt to answer the question "Will NIST's PQC Standardization yield feasible post-quantum secure digital signature drop-in replacements for all applications?"".

"PQC implementations on SoCs (system-on-chips), ASICs (application-specific integrated circuits), and FPGAs (field-programmable gate arrays) are active fields of research [12–16] and have yielded promising results."

"A search on related research implementations of PQC algorithms on chip cards yielded no signature examples and two encryption examples [17, 18]."

"To answer the question: Q: Will NIST's PQC Standardization yield feasible post-quantum secure digital signature drop-in replacements for all applications?"

"Work on measuring the NIST PQC Standardization round 3 candidates on chip card platforms will be an important part of PQC research."

"We expect that improvements in these two areas will make Picnic a strong signature candidate for chip card applications."

Breaking Barriers in Conventional Cryptography by Integrating with Quantum Key Distribution [19]

This is a machine-generated summary of:

Ahilan, A.; Jeyam, A.: Breaking Barriers in Conventional Cryptography by Integrating with Quantum Key Distribution [19].

Published in: Wireless Personal Communications (2022).

Link to original: https://doi.org/10.1007/s11277-022-10110-8

If you want to cite the papers, please refer to the original.

For technical reasons we could not place the page where the original quote is coming from.

Abstract-Summary

"Security techniques based on conventional cryptography assume keys are disseminated prior to secure communications in traditional security mechanisms."

"Quantum cryptography is a method of securely encrypting information sent between parties while also detecting intruders attempting to listen in on the discussion."

"We discuss the Quantum key distribution (BB84 protocol) and how when integrated with conventional cryptography algorithms it increases security in data transmission to a remarkably high level."

"We also compare the cryptography algorithms for different file sizes and measure their performance by calculating the Encryption,decryption,throughput and Avalanche effect of the algorithms with and without QKD."

Introduction

"Quantum cryptography varies from conventional cryptography systems in that physics plays a fundamental role in its security concept whereas in conventional cryptography, mathematics plays a key role."

"Hackers that use quantum computing as part of their attack arsenal will be able to quickly decipher today's encryption methods."

"While Quantum cryptography makes use of quantum mechanics to safeguard key exchanges."

"Weak random key generators, advancements in CPU power, new attack techniques, and the development of quantum computers all challenge the security of traditional encryption."

"Data that is encrypted today can be intercepted and saved in the future for decryption by quantum computers."

"The major contribution of this paper is given as follows; QKD is integrated with classical cryptography algorithms it increases the security in transmitting data to a high level Quantum key distribution (QKD) allows users to safely exchange conventional keys, which can then be utilized for secure communication."

Quantum Key Distribution (QKD) Using BB84 Protocol

"Alice begins the communication by sending Bob a secret key that will be used to encrypt the data."

"Bob must select whether to read each photon bit using his rectilinear or diagonal polarizer when he receives the photon key: sometimes the correct polarizer will be chosen by Bob, and at other times he will choose the incorrect one."

"To obtain the key correctly, she'd need to know the filters which were used by Alice and Bob before sending the photons [20]."

"Alice and Bob converse while exchanging quantum keys via both the public/classical channel and the quantum channel."

"Following the bit error estimation, Alice and Bob use key distillation procedures."

"Alice and Bob filtered the key into blocks and checked the parity."

"Because Eve may have obtained critical data about the secret key, Alice and Bob must enhance their secrecy (to prevent eavesdropping during bit error estimation in the quantum channel and key reconciliation in the public channel)."

Comparison of Quantum Cryptography Protocols

"To illustrate the reliability of QKD protocols [21], several protocols were established."

"QKD protocols can be implemented using an existing security system."

"These QKD protocols were created in a variety of ways, and some of them required specific hardware."

Other Methods for Secret Key Exchange

"The quantum key distribution (QKD) is a perfect illustration of the value of quantum effects in the creation of provably secure techniques for exchanging secret keys in cryptography."

"Bob's photon will be projected to the appropriate polarisation state in accordance with the measurement result from Alice."

"Bob's measurement will be entirely interrelated with Alice's if bob reads his photon on the same bases as Alice."

"Between Alice and Bob, the EPR source can be put."

"Each EPR pair sends one photon to Alice and the other to Bob."

"Alice and Bob choose rectilinear or diagonal measurement bases for each incoming photon at random and independently."

"When they've finished, only the photon pairs that Alice and Bob have measured will be kept when they compare their measurement bases."

"That the polarization of each photon is unknown before Alice and Bob do the measurement."

Results and Discussion

"The BB84 Protocol was implemented using the Python programming language since it offers the freedom to select the necessary modules for the creation of the code and simulation of the protocol."

"To implement the code, support packages such as pool, random, and certain system packages were needed."

"This study compares the performance of three of the most used algorithms, including 3DES, AES, and Blowfish, using varied processing settings to examine different data file sizes ranging from 500 to 3500 KB."

"The time it takes to encrypt a message is measured in milliseconds and is dependent on the size of the key and the size of the text."

"It shows that when combined with QKD, it outperforms the classical cryptography algorithms without QKD for file size = 3500 Kb."

"The Execution time is evaluated against the file size and the file size is varied from 100 Kb to 35oo Kb."

Conclusion

"This paper discusses Quantum Cryptography and how quantum cryptography is more efficient when compared with conventional cryptography algorithms and shows when QKD is integrated with classical cryptography algorithms it increases the security in transmitting data to a high level."

"It can be seen that QKD with classical cryptography decreases the encryption and decryption time and increases the throughput for file sizes ranging from 500 to 3500 Kb."

"The chance of detecting eavesdroppers in quantum cryptography is high compared to conventional cryptography algorithms."

A Quantum Resistant Anonymous Poxy Signature Scheme [22]

This is a machine-generated summary of:

RAWAL, SWATI; PADHYE, SAHADEO: A Quantum Resistant Anonymous Poxy Signature Scheme [22].

Published in: Sādhanā (2022).

Link to original: https://doi.org/10.1007/s12046-022-01808-3

If you want to cite the papers, please refer to the original.

For technical reasons we could not place the page where the original quote is coming from.

Abstract-Summary

"Occasionally anonymity of proxy signers is required so that no one can reveal their identity, including the original signer."

"The paper thus proposes a quantum-safe anonymous proxy signature, which provides anonymity to the proxy signers based on the worst-case hardness of lattice problems."

Introduction

"Jiang et al. [23] introduced the first lattice-based proxy signature, offering a quantum secure alternative."

"Wang et al. [24] proposed a proxy signature scheme using the bonsai trees concept."

"Xia et al. [25] also gave a proxy signature using the bonsai trees, which security is based on the hardness of average-case SIS and ISIS problems."

"The lattice-based ID-based proxy signature was first proposed by Zhang et al. [26] using the bonsai tree, which is proved to have a unforgeability proxy key, revocability of proxy signature as well as existential unforgeable."

"Kim et al. [27] constructed the first ID-based proxy signature scheme with proxy protection in the random oracle model."

"In 2018, Faguo Wu et al. proposed an ID-based proxy signature [28], but based on NTRU lattice in the random oracle model."

Anonymous Proxy Signature: Formal Structure and Security Requirements

"KeyGen: On receiving the system parameters as input, the algorithm generates the public/private keys of original signer and the proxy signers."

"Type 1: The adversary here have the public and private key pair of original signer and public keys of proxy signers."

"Type 2: The adversary here have the public keys of original and proxy signers, and private keys of some of the proxy signers."

"Type 3: The adversary only have access to the public keys of both original and proxy signer."

"Existential unforgeability adverse to Type 1 adversary: The anonymous proxy signature is said to unforgeable against Type 1 probabilistic polynomial time adversary if the probability of successfully executing the following game is non-negligible."

"Existential unforgeability adverse to Type 2 adversary: The anonymous proxy signature is said to unforgeable against probabilistic polynomial time Type 2 adversary if the probability of successfully executing the following game is non-negligible."

Security Analysis

"The unforgeability of the scheme is proved under the hardness of the short integer solution problem, for all possible types of adversaries."

"The proposed construction is secure against Type 1 adversary under the event that underlying hash H is collision-resistant, and SIS is hard."

"Type 1 adversary can lead in two different ways."

"Our scheme has no overwhelming probability in winning the above game, making the scheme to attain anonymity under the hardness of SIS problem."

Conclusion

"The paper introduces a provable secure lattice-based anonymous proxy signature scheme."

"The proposed signature works when one out of a number of proxy signers sign messages with unconditional anonymity on behalf of the original signer."

"This scheme is the first lattice-based anonymous proxy signature scheme to the best of our knowledge."

"We have proved that our scheme is anonymous and existential unforgeable against adaptive chosen-message attack based on the hardness of the short integer solution problem in lattices for the security analysis."

Post-Quantum Hybrid Key Exchange: A Systematic Mapping Study [29]

This is a machine-generated summary of:

Giron, Alexandre Augusto; Custódio, Ricardo; Rodríguez-Henríquez, Francisco: Post-quantum hybrid key exchange: a systematic mapping study [29].

Published in: Journal of Cryptographic Engineering (2022).

Link to original: https://doi.org/10.1007/s13389-022-00288-9

If you want to cite the papers, please refer to the original.

For technical reasons we could not place the page where the original quote is coming from.

Abstract-Summary

"Since it was apparent that large-scale quantum computers can comfortably break most commonly used cryptographic protocols, proposals of quantum-safe KEX began to appear."

"Some of these proposals combine well-known classical cryptographic key exchange protocols with novel post-quantum schemes."

"We present a systematic mapping study of post-quantum hybrid KEXs, given insights into their characteristics, terminology, efficiency, security and open design challenges."

"These features indicate that hybrid KEX will shortly become an important building block for secure network communications, even in a worst-case scenario where large-scale quantum computers are prevalent."

Introduction

"To the apparent advantage of interoperability, some authors point out that hybrid systems are interesting, as security can be based on two different algorithms: one pre-quantum and one post-quantum."

"For both KEX and authentication, the purpose of Post-Quantum Hybrids is to negotiate two or more algorithms, for example, a classic and a post-quantum, and then proceed using the two combined."

"Along these lines, Schank and others [30] stated the transitional security property that hybrid KEX can have, namely, when it provides pre-quantum authentication and post-quantum confidentiality."

"There is no systematic study addressing Post-Quantum hybrid KEX schemes."

"This contribution is also expected from this paper, i.e., to provide precise terminology that aids to have a better understanding of the hybrid Post-Quantum KEX research."

Background

"A Key-Establishment Method is a synonym of KEX [31], both meaning a method that enables two parties to share a cryptographic key."

"As soon as the communicating parties establish the shared secret, a Key-Derivation Method is often used."

"It is important to mention that this definition does not refer explicitly to Quantum Key Distribution (QKD), which could be part of a Hybrid KEX."

"As the short-term scenario is being considered in this paper, in which the communicating parties do not have access to a QKD infrastructure, only pre- and post-quantum KEX methods compose this hybrid form."

"Several proposals for post-quantum KEX are designed as encapsulation mechanisms (KEMs), not as key-agreement."

"Although that it is not necessarily part of a KEX design, key-derivation must preserve security properties of the KEX, such as the key indistinguishability [32]: the shared secret of the KEX should be indistinguishable from one chosen randomly."

Search Methodology

"This systematic mapping study aims to identify and classify Hybrid KEXs that have been proposed in the literature."

"In order to select the studies that also address key-derivation, the related terms are included in the "Comparison" part."

"Each preliminary study that was selected was also subject to snowballing, both in backward snowballing (i.e., searching the references) and in forwarding snowballing (i.e., searching in the citations of the primary study), in an iterative process."

"The selection process resulted in an initial set of 23 primary studies: 16 articles, 4 PhD/Master theses and 3 technical reports."

"In this systematic mapping 29 primary studies were selected for analysis."

"There are 58 authors and 29 selected primary studies."

The Hybrid Design

"Paquin and others [33] proposed a hybrid KEX for TLS, where the cryptographic data is concatenated (in the TLS keyshare)."

"The final approach is a Nested dual-PRF combiner to derive a key K, which results in the following construction:The authors proposed a model of this combiner be applied in TLS 1.3 [34]."

"Giacon and others also proposed split-key PRFs for KEM combiners [35], proved to be secure under the Chosen-Ciphertext Attack (CCA), under the assumption that at least one KEM of the hybrid setting is CCA-secure."

"A hybrid KEX that employs an intermediary combiner delegates the key-derivation for an external KDF."

"Primary studies that evaluate hybrid KEX in TLS protocol used the HKDF "as-is."".

"Ghosh and Kate [36] proposed an one-way Authenticated Key Exchange (1 W-AKE) hybrid protocol, where the core combining function was an XOR operation between the shared secrets."

Open Challenges

"Crockett and others [37] state that it is important to avoid "wasting bytes" in sending KEX data when deploying hybrids over network protocols."

"Hybrid Certificates can be built using dual signatures (i.e., a classical and a PQC algorithm) or using KEMs, such as in the KEMTLS approach [38]."

"The focus on hybrid design for KEX (and for Digital Signatures) puts forward the research on combiners and their security."

"Although an adversary could, in theory, employ side-channel attacks individually and then break the hybrid scheme, we were not able to find any primary study addressing this problem."

"The evaluation and analysis of different hybrid KEX designs help developers and engineers to understand and design their protocol accordingly."

"Some designs are available already, e.g., the OQS project [39], and it is expected that more evaluations and analysis will contribute to the adoption of Hybrid PQC shortly."

Final Remarks

"The popularization of hybrid KEXs is aimed as a way to ease the transition to post-quantum cryptography."

"A classification was provided for the key aspects of the design of hybrid KEXs: how to convey and combine cryptographic data; and approaches used so far for key-derivation."

"More research is recommended in order to fully understand the derivation aspect of hybrid KEX."

"Hybrid constructions for the post-quantum transition process may be a temporary approach."

"Hybrid approaches will still be needed until the confidence in the security of the post-quantum cryptography is well established."

Supplementary Information

"The supplementary information of this study is available at https://disk.yandex.com/d/CzSDnR6Ab16_KA."

"This supplementary material is provided for verification and reproducibility purposes."

Learning Classical Readout Quantum PUFs Based on Single-Qubit Gates [40]

This is a machine-generated summary of:

Pirnay, Niklas; Pappa, Anna; Seifert, Jean-Pierre: Learning classical readout quantum PUFs based on single-qubit gates [40].

Published in: Quantum Machine Intelligence (2022).

Link to original: https://doi.org/10.1007/s42484-022-00073-1

If you want to cite the papers, please refer to the original.

For technical reasons we could not place the page where the original quote is coming from.

Abstract-Summary

"Some of these constructions apply single-qubit gates in order to provide a secure fingerprint of the quantum device."

"We formalize the class of classical readout quantum PUFs (CR-QPUFs) using the statistical query (SQ) model and explicitly show insufficient security for CR-QPUFs based on single-qubit rotation gates, when the adversary has SQ access to the CR-QPUF."

"We demonstrate how a malicious party can learn the CR-QPUF characteristics and forge the signature of a quantum device through a modelling attack using a simple regression of low-degree polynomials."

"We thoroughly discuss the prospects and problems of CR-QPUFs where quantum device imperfections are used as a secure fingerprint."

Introduction

"The recent work by Phalak et al. [41] proposes a QPUF, where the authentication protocol requires only classical communication and no quantum memory, and can be implemented on gate-based quantum computers to fingerprint these devices."

"We adapt the QPUF framework of Doosti et al. [42] in order to address the case of classical readout quantum PUFs (CR-QPUFs) where verifier and prover communicate classically to authenticate a quantum device."

"Even though it has been shown that restrictions to the SQ model can be used to obtain unlearnability results (Hinsche et al. [43]; Gollakota and Liang, [44]; Kearns, [45]), we present a successful learning attack on the Hadamard CR-QPUF in the SQ model."

"We show that an attacker is able to model and predict the Hadamard CR-QPUF characteristics and hence forge the quantum device fingerprint using machine learning."

"It is possible that in the NISQ era, constructing CR-QPUFs in the statistical query (SQ) model might provide security guarantees against learning attacks."

Classical-Readout QPUFs

"This enables her to build a secret CRP database by repeatedly querying the QPUF with randomly chosen challenges and storing the challenges and respective responses in the database."

"At a later stage, when Bob wants to prove possession of the QPUF to Alice, he will need to provide the correct corresponding response to a CRP chosen randomly by Alice from her database."

"If Alice wants to authenticate the QC at a later point when she no longer has physical access to the hardware, she queries the QPUF again with challenges from her CRP database and compares the respective responses."

"Prior work of Škorić [46] introduced the concept of a quantum-readout PUF (QR-PUF), where challenges and responses are communicated using a quantum channel, thereby presenting a protocol to authenticate a QR-PUF without the need to rely on a trusted readout device."

The Hadamard CR-QPUF

"The Hadamard CR-QPUF introduced by Phalak et al. [41] aims at robustly authenticating a quantum computer using device-specific qubit imperfections via a classical communication channel."

"The CR-QPUF challenges are given by a parameterized quantum circuit that includes parameterized single-qubit rotations and Hadamard gates."

"A natural extension of the Hadamard CR-QPUF as proposed by Phalak et al. [41] is to increase the number of rotations applied to the qubits to rotation chains."

"In our modelling attack on the Hadamard CR-QPUF, we also consider such natural extensions and show that the resulting responses can be learned using a simplified model."

Attacker Capabilities

"One such example is when a customer is assigned by a malicious actor, to cheaper and lower-grade quantum computing hardware than what they had originally agreed upon."

"This includes that the cloud provider is malicious and aims at assigning the customer to lower-grade hardware."

"This is well in the threat model of a malicious quantum cloud provider."

"As we will show in the following section, such a malicious actor that can reroute user circuits to lower-grade remains completely undetected by the Hadamard CR-QPUF secure provisioning protocol."

The Modelling Attack

"The attack on the Hadamard CR-QPUF secure provisioning scheme is carried out in two phases, the learning phase and the attack phase."

"One important step in the attack is based on the observation that the Hadamard CR-QPUF does not entangle the single qubits and thus creates the opportunity to learn the characteristics of the qubits individually."

"We describe the learning phase, where the design shortcomings of the Hadamard CR-QPUF are exploited to learn a model of the CR-QPUF, and the attack phase, where the actual attack using the learned models is carried out."

Results

"Almost all of the predicted responses get accepted."

"We were able to model the Hadamard CR-QPUF behaviour and successfully predicted responses to unknown challenges."

"This showcases the predictability and thus the insecurity of the Hadamard CR-QPUF."

"In the attack phase, we aim at predicting responses to unknown challenges comprising chains of rotations."

"Again almost all of the predicted responses will get accepted by the protocol, showcasing the insecurity of these extensions of the Hadamard CR-QPUF in the SQ model."

Discussion

"In the context of noisy quantum devices, this situation poses a very interesting perspective: since NISQ devices are inherently noisy, can one leverage the restriction to the SQ model to derive provable security against learning attacks?"

"Can one construct a CR-QPUF, such that one can show that learning the CR-QPUF in the SQ model implies learning an unknown stabilizer state in the SQ model or an unknown local quantum circuit in the SQ model?"

"CR-QPUF proposals could be checked against classical and quantum learning attacks, providing evidence to the question whether NISQ computations form a low-complexity class of algorithms whose output can be learned using low-degree polynomials."

"As we have found in the insecurity of the Hadamard CR-QPUF, the responses can be learned using low-degree polynomials, which shows us that the class of challenge unitaries needs to be designed very carefully."

Future Work

"Going forward, analyzing and categorizing the device imperfections and their influence on degenerating the Born distribution output would serve a better understanding of CR-QPUFs."

"Can device imperfections be leveraged such that the mutation of the output distribution of local quantum circuits is not learnable in the SQ model?"

"This poses an excellent playground to gain further insight to CR-QPUFs based on quantum device imperfections."

Complexity Limitations on One-Turn Quantum Refereed Games [47]

This is a machine-generated summary of:

Ghosh, Soumik; Watrous, John: Complexity Limitations on One-turn Quantum Refereed Games [47].

Published in: Theory of Computing Systems (2022).

Link to original: https://doi.org/10.1007/s00224-022-10105-9

If you want to cite the papers, please refer to the original.

For technical reasons we could not place the page where the original quote is coming from.

Abstract-Summary

"This paper studies complexity theoretic aspects of quantum refereed games, which are abstract games between two competing players that send quantum states to a referee, who performs an efficiently implementable joint measurement on the two states to determine which of the player wins."

"The complexity class QRG(1) contains those decision problems for which one of the players can always win with high probability on yes-instances and the other player can always win with high probability on no-instances, regardless of the opposing player's strategy."

"If one of the players is limited to sending a classical (probabilistic) state rather than a quantum state, the resulting complexity class CQRG(1) is contained in ∃·PP (the nondeterministic polynomial-time operator applied to PP); while if both players send quantum states but the referee is forced to measure one of the states first, and incorporates the classical outcome of this measurement into a measurement of the second state, the resulting class MQRG(1) is contained in P ·PP (the unbounded-error probabilistic polynomial-time operator applied to PP)."

Introduction

"Two complexity classes are defined—RG(1) in the classical setting and QRG(1) in the quantum setting—consisting of all promise problems $A = (A_{yes}, A_{no})$ for which there exists a game (either classical or quantum, respectively) such that Alice can win with high probability on inputs $x \in A_{yes}$ and Bob can win with high probability on inputs $x \in A_{no}$, regardless of the other player's behavior."

"This class is most typically defined in terms of quantifiers that suggest games in which Alice and Bob choose polynomial-length strings (as opposed to probability distributions of strings) to send to the referee, but the class does not change if one adopts a bounded-error definition in which Alice and Bob are allowed to make use of randomness [48]."

"Gutoski and Wu [49] proved that even the class QRG(2), which is analogous to QRG(1) except that the referee first sends a message to Alice and Bob and then receives responses from them, is contained in PSPACE."

Preliminaries

"Unless it is explicitly indicated otherwise, the input of a given polynomially bounded function p is assumed to be the natural number $|x|$, for whatever input string $x \in \Sigma^*$ is being considered at that moment."

"By the assumptions on our pairing function described above, it is the case that $|\langle x, y\rangle|$ depends only on $|x|$ and $|y|$, and therefore there exists a (necessarily polynomially bounded) function r such that $r(|x|) = p(|x|) + q(|\langle x, y\rangle|)$ for all $x \in\Sigma^*$ and $y \in\Sigma^p$."

"We choose to use the standard, general model of quantum information based on density operators and quantum channels, as opposed to the restricted model of pure state vectors and unitary operations, when discussing quantum circuits."

"If a quantum circuit were to be represented in a standard way as a directed acyclic graph, its size would simply be the number of vertices, including a vertex for each input and output qubit, of the corresponding graph."

Complexity Classes for One-Turn Quantum Refereed Games

"R_x takes an n-qubit register A and an m-qubit register B as input, measures each qubit of A with respect to the standard basis, leaving it in a classical state, and then runs the circuit Q_x on the pair (A,B), producing a single output qubit."

"For any density operator ρ that Alice might choose to play, the state of A that is input into Q_x takes the form for some probability vector p over n-bit strings, and therefore the state that is plugged into the top n qubits of the circuit Q_x represents a classical state."

"The register Y is then measured with respect to the standard basis, so that it then contains a classical state, and finally a quantum circuit Q_x is applied to the pair (Y,B), yielding a single qubit."

"An MQRG(1) referee measures Alice's qubits with respect to a general, efficiently implementable measurement, which yields a k-bit classical outcome, which is then plugged into Q_x along with Bob's quantum state."

Upper-Bound on CQRG(1)

"The proof represents a fairly direct application of the Althöfer–Lipton–Young [50, 51] technique, although (as was suggested above) the quantum setting places a new demand on this technique that requires the use of a tail bound on sums of matrix-valued random variables."

"Consider first the situation that Alice plays deterministically, sending a string y $\in\Sigma^n$ to the referee, so that $\rho = |y\rangle\langle y|$."

"Having selected a state ρ representing Alice's play, we are effectively left with a binary-valued measurement being performed on the state sent to the referee by Bob."

"A large minimum eigenvalue means that Alice has managed to force the outcome 1 to appear, regardless of what state Bob plays, whereas a small minimum eigenvalue means that Bob has at least one choice of a state that causes the outcome 1 to appear with small probability."

Upper-Bound on MQRG(1)

"By Lemma 7 there exists a polynomially bounded function r and GapP functions f_0 and f_1 such that for all x $\in\Sigma^*$, z, w $\in\Sigma^n$, and u, v $\in\Sigma^k$."

"Define for all x $\in\Sigma^*$, z, w $\in\Sigma^n$, and u $\in\Sigma^k$."

"Define for all x $\in\Sigma^*$ and z, w $\in\Sigma^n$."

"We observe that for all x $\in\Sigma^*$ and z, w $\in\Sigma^n$, where Now let us consider the cases x $\in A_{yes}$ and x $\in A_{no}$."

"As in the proof of Theorem 14, define an operator for each x $\in\Sigma^*$ and y $\in\Sigma^k$."

"To prove that the promise problem A is contained in QMA ·PP, it suffices to prove two things: The proof of completeness follows a similar argument to the proof of Theorem 14."

Conclusion

"A diagram illustrating the containments is provided."

"We did not succeed in this endeavor, and so we leave this as an open question."

"Observing that the containments we prove establish that CQRG(1) and MQRG(1) are contained in the counting hierarchy, we ask specifically: is QRG(1) also contained in the counting hierarchy?"

Lightweight Two-Factor-Based User Authentication Protocol for IoT-Enabled Healthcare Ecosystem in Quantum Computing [52]

This is a machine-generated summary of:

Al-saggaf, Alawi A.; Sheltami, Tarek; Alkhzaimi, Hoda; Ahmed, Gamil: Lightweight Two-Factor-Based User Authentication Protocol for IoT-Enabled Healthcare Ecosystem in Quantum Computing [52].

Published in: Arabian Journal for Science and Engineering (2022).

Link to original: https://doi.org/10.1007/s13369-022-07235-0

If you want to cite the papers, please refer to the original.

For technical reasons we could not place the page where the original quote is coming from.

Abstract-Summary

"This paper presents a new two-factor-based user authentication protocol for the IoT-enabled healthcare ecosystem in post-quantum computing environments using the PQFC scheme."

"The proposed protocol is proved to be secure using random oracle model."

"The functionality and security of the proposed protocol are analyzed, showing that memoryless-effortless, user anonymity, mutual authentication, and resistance to biometric templates tampering and stolen attacks, stolen smart card attack, privileged interior attack are fulfilled."

"The results demonstrate that the proposed protocol is more efficient than Mukherjee and others, Chaudhary and others, and Gupta and others protocols."

Introduction

"The most significant threats that IoT-enabled healthcare poses are data security and privacy."

"The IoT-enabled healthcare security is mainly for secure health records, communication, and user authentication."

"User authentication is a keystone in IoT-enabled healthcare security, which plays a crucial role in establishing trust between IoT healthcare users and devices and preventing attacks [53]."

"Post-quantum cryptography primitives are a promising technique for securing communications between IoT users and devices."

"To tackle issues with IoT-enabled healthcare ecosystems, we propose a new lightweight two-factor user authentication protocol for the IoT-enabled healthcare ecosystem based on the security of PQFC scheme."

"A new lightweight two-factor user authentication protocol for the IoT-enabled healthcare ecosystem using a post-quantum fuzzy commitment scheme."

"The security and performance analysis shows that the proposed protocol is suitable for application in an IoT-enabled healthcare environment in comparison with the other existing competitive protocols."

Related Work

"Many authentication protocols for secure communication between IoT users and devices in IoT environments have been proposed."

"There are also less efficient and secure authentication protocols [54–57], which are based on traditional biometric template protection."

"The security of their protocol is based on the lattice-based homomorphic encryption."

"Sahu and others [58] presented a lightweight multi-party authentication and key-establishment protocol in an IoT-based e-Healthcare service access network using lattice identity-based encryption."

"Gupta and others [59] presented a lattice-based authentication and access control protocol for IoT-based healthcare."

"All the aforementioned authentication protocols for IoT environments are relying solely on the password, which is falling apart if the password is not kept secure."

"The rapid development of emerging technologies such as IoT, cloud computing, blockchain, quantum computing, and e-services makes the current research on user authentication protocols based on post-quantum cryptography urgent."

Preliminaries

"We now give definitions of well-known lattice computational problems used to construct lattice-based cryptography primitives."

"LP7: Learn with Error (LWE) problem: We briefly describe the Learn with Error (LWE) that used to construct an efficient lattice-based cryptography."

"Regev [60] introduced a reduction from worst-case lattice problems such as GAPSVP and SIVP to a learning with error problems."

The Biometric-Based PQFC Authentication System

"We briefly describe the biometric-based PQFC authentication system [61], which is relies on the worst-case hardness shortest vector problem (SVP) of lattice cryptography."

"Let us now describe the construction of the biometric-based PQFC authentication system which consists of two main stages: enrollment and verification."

"The process of the system is described below: Positive integers m, n, and p (prime number) are chosen randomly."

"If the matching score is within the system threshold, then the user is authenticated."

Lightweight Two-Factor User Authentication Protocol for the IoT-Enabled Healthcare

"The proposed protocol comprises four phases, namely the registration phase, the login phase, the authentication phase, and the biometric renewable phase."

"The patient's medical data are collected and measured using smart devices implanted with the body of the patient."

"Details of the steps of these phases are described below."

Security Analysis

"A formal security analysis of the proposed protocol is given using the random oracle model (ROM)."

"We simulate two random oracle model."

"We are computing the probability of the distribution: Then, we are ready to estimate the probability between the two distributions."

Security and Functionality Features

"F2: Tampering with stored biometric templates attack: This property applies when an attacker gets access to the system database or the token, temporarily or permanently cannot modify the template in the system database/token to gain server authentication."

"In the proposed protocol, the attacker needs to break the SVP problem to obtain the biometric reference template."

"F3: Biometric template thefts resistant: This property applies to an attacker that gets access to the database system or token and obtain the user's biometric template; she/he can use it for other purposes."

"The proposed protocol offers an opportunity for the user to hide her/his biometric template from privileged insiders in the registration phase by allowing her/him to send it to authentication server in encrypted format, which will prevent an inside attacker from getting it."

"By this definition, the proposed authentication protocol is memoryless-effortless F9: User anonymity: An important security property of authentication protocol for IoT applications is the confidentiality of the user's identity."

Performance Analysis

"We evaluate the performance of our protocol based on the following metrics: the storage requirements, communication costs, and computational complexities."

"We have compared the proposed protocol with the recent related protocols for IoT systems [59, 62]."

"The total execution time for the proposed protocol is 20.0437 ms."

"The storage requirements of our protocol and the related protocols [59, 62] are computed."

Conclusion

"The proposed protocol achieved the following functionality and security properties: memoryless-effortless, user anonymity, mutual authentication, and resistance to tampering and stolen of biometric template, stolen smart card, privileged interior attacks."

"The proposed protocol was evaluated in terms of the performance metrics: storage requirement, computation and communication."

"The overhead of the computational costs of our protocol becomes larger naturally since the proposed protocol exploits these computations to provide several significant security and functionality properties."

"The overall performance demonstrates that the proposed protocol is suitable for the Internet of Things applications."

A Quantum Secure ID-Based Cryptographic Encryption Based on Group Rings [63]

This is a machine-generated summary of:

Mittal, Gaurav; Kumar, Sunil; Kumar, Sandeep: A quantum secure ID-based cryptographic encryption based on group rings [63].

Published in: Sādhanā (2022).
Link to original: https://doi.org/10.1007/s12046-022-01806-5

If you want to cite the papers, please refer to the original.

For technical reasons we could not place the page where the original quote is coming from.

Abstract-Summary

"The security of most of the ID-based encryption schemes is based directly or indirectly on solving integer factorization problem, Elliptic curve discrete logarithm problem or discrete logarithm problem."

"In this paper we propose a secure ID-based encryption scheme whose security depends on the newly discovered hard problems in the algebraic structure of group rings."

"We show that the proposed scheme is IND-ID-CPA secure and safe against the chosen ciphertext attack."

"We also comment on the IND-ID-CCA security of the proposed scheme."

Introduction

"There is a demand of novel ID-based encryption schemes whose security depends on other hard problems that are safe against attacks on quantum computers."

"There are several ID-based encryption schemes proposed in the literature that are claimed to be quantum secure."

"We continue in the direction of construction of quantum safe ID-based encryption schemes and consider an algebraic structure (i.e., a set with two binary operations satisfying few properties) that is source of many hard problems known as group rings (see [64] for an excellent survey on group rings)."

"We utilize the hard problems of group rings to construct a novel ID-based encryption scheme."

"We propose a secure (even in quantum sense) and novel ID-based encryption scheme that is based on the structure of group rings."

Preliminaries

"We discuss some basic definitions that are pre-requisite in our work."

"This section contains definitions from [64–67]."

"A binary operation is a calculation that combines two elements to produce another element."

Identity-Based Encryption Scheme

"We present our identity-based encryption scheme, called IBES by describing the following four algorithms: Parameters-setup, Extract, Encrypt and Decrypt."

"This can be done by generating the known type units and take their product or taking the power of a known type unit."

"By a known type unit, we mean to say a unit that can be generated with the help of a certain formula, e.g. Bicyclic units, Bass cyclic units etc., [64]."

Security

"The proposition 4.1 assures that the IND-ID-CPA security of the scheme IBES depends on the IND-CPA security of the scheme PES."

"We check for the IND-CPA security of the scheme PES."

"It is worth studying the IND-CPA security of PES scheme by assuming that ECDLP is solvable."

"We study the IND-CPA security of the scheme PES."

"In all, we have shown that if DDHP is a hard problem in group rings, then IBES is IND-CPA secure."

"We want to mention that the hardness of DLPGR in group rings is not only due to the fact that there is currently no polynomial time algorithm to solve it, but also that it is well studied by the mathematically community."

"Theorem 4.1 implies that our novel IBES scheme is IND-ID-CPA secure."

Example

"We discuss a toy example to demonstrate our scheme IBES."

"One can efficiently compute the Weil pairing by using the Miller's algorithm [68]."

"Let the hash functions be and we can take the standard hash function, e.g., SHA-256 and truncate the 256 bit hash value to obtain the desired number of bits."

Computational Cost and Parameters Size

"We discuss the computational cost of encrypting and decrypting a message through IBES and the size of parameters required to securely implement IBES."

"One can use Weil pairing that is a well known Bilinear pairing."

"We discuss how to select various parameters as well as hash functions required in IBES."

"Further, one may use Weil pairing or Tate pairing as bilinear pairing, since both can be efficiently computed [66]."

"This completes the description of hash functions as well as parameters size for IBES."

Comparison with Other Post Quantum Cryptographic Primitives

"We discuss about the various Post quantum cryptographic primitives and compare the group rings based hard problems with other hard problems."

"The problem of constructing an isogeny is hard but no security reduction to a NP-hard problem is known."

"The problems such as DLPGR and ICPGR are recently discovered hard problems [69], whereas the hard problems in lattices and coding theory etc., are well studied."

"The hardness of DLPGR and ICPGR can be comparable to that of isogeny problem in supersingular elliptic curve cryptography."

"The advantage of utilizing hard problems such as DLPGR and ICPGR is that they provide a security equivalent to 112 bits or more with the small size parameters as compare to other group based schemes."

"It is worth to mention that it is an open problem to deduce whether or not the problems DLPGR and ICPGR are NP-hard."

Conclusion

"We have shown the IND-ID-CPA security of the scheme by utilizing the hardness of hard problems in group rings."

"Further, we have also discussed about the various transformations that can turn our IND-ID-CPA secure scheme to a IND-ID-CCA secure, even in a post-quantum setting."

"We want to emphasize the fact that our scheme requires considerably small sizes of the parameters in order to provide the same level of security as provided by the current secure implementations of ECDLP and RSA (or DLP)."

"Due to the small size of the involved parameters and strong security, we conclude that our proposed scheme can be employed in the near future in place of the existing schemes."

AKAME: A Post-Quantum Authenticated Key-Agreement and Message Encryption Scheme Based on Ring-LWE [70]

This is a machine-generated summary of:

Choudhary, Simran; Gupta, Anil: AKAME: A post-quantum authenticated key-agreement and message encryption scheme based on ring-LWE [70].

Published in: International Journal of Information Technology (2022).

Link to original: https://doi.org/10.1007/s41870-022-00888-y

If you want to cite the papers, please refer to the original.

For technical reasons we could not place the page where the original quote is coming from.

Abstract-Summary

"The existing post-quantum cryptographic primitiveshave longer key sizes and high computational time."

"This paper presents an authenticated quantum secure key-agreement and message encryption (AKAME) scheme with optimal key-sizes and decryption time."

"The proposed scheme is capable of providing both 2^{118} or 2^{254} classical security and 2^{102} or 2^{241} quantum security."

"Further, the paper presents comparative analysis between existing schemes and presented scheme."

Introduction

"Since as per Peter Shor's work, this may lead to decryption of most of the world's secure data encrypted by cryptographic primitives based on integer factorization and discrete logarithm integer problems [71]."

"This homogeneity in lattices results complexity in lattice-based constructions making them quantum safe [72]."

"Lattice-based primitives have comparatively efficient implementations than classical cryptosystem [73]."

"LBC is versatile as it comprises of various cryptographic primitives like hash functions, public-key encryption schemes, digital signatures, identity-based encryption, and fully-homomorphic encryption schemes etc. [74, 75]."

"Many lattice-based cryptographic primitives were proposed for public key exchange (KEX) and key encapsulation mechanism (KEM) to NIST call for standardization of post-quantum cryptography [73, 74]."

"This paper presents a novel and robust lattice-based scheme AKAME for quantum-safe authenticated key-agreement and message encryption cryptographic primitive without any reconciliation technique."

"The proposed scheme is tested against brute-force attack and lattice-based attacks [76, 77]."

Theoretical Basis

"Beside these, Learning with Errors (LWE) and Ring Learning with Errors (RLWE) are popular lattice-based problems [78, 79]."

"This has inspired the usage of RLWE which allows achieving the same security as LWE based schemes with a smaller key size of ($\sqrt{n}$) [80]."

"The RLWE problem is proved to be hard using a quantum reduction from worst-case approximate SVP on ideal lattices to the search version of RLWE [81–84]."

"The first RLWE based key-exchange protocol was introduced by Regev and later it was improved by Ding [85]."

"In 2016, NewHope scheme was introduced which uses reconciliation scheme similar to Peikert for key agreement [86]."

"The RLWE based lattice cryptosystem are feasible and practical, however the existing schemes have certain limitation like failure of key-agreement, large key sizes, and encryption/decryption time [87]."

"This paper addresses these limitations and present an efficient RLWE based key-agreement and message encryption AKAME scheme with zero failure of key-agreement."

AKAME Scheme

"The AKAME scheme is given by three algorithms namely, AKAME. KeyGeneration, AKAME.MsgEncryption, AKAME.MsgDecryption, defined respectively in Algorithm 1, Algorithm 2, and Algorithm 3."

"The scheme starts with AKAME.KeyGeneration algorithm."

"Most significant 256-bits of common secret is used as key for encrypting the message of size 256-bits by XOR operation."

"The Algorithm 2 AKAME.KeyEncryption presents the entire procedure followed by second party."

"The first party performs the computation of common secret key at its own end."

"The key-agreement and message encryption procedure is outlined in Algorithm 3 AKAME.MsgDecryption."

"The proposed AKAME scheme performs the key-agreement without sharing any reconciliation information."

"Further, matching the hash digest ensure successful key-agreement between two parties and prevents wrong decryption of message too."

Implementation & Results

"The proposed RLWE-based AKAME scheme is implemented in C language."

"Several experiments are performed using different parameters and the performance of the software implementation of proposed scheme is evaluated on Dell laptop with an Intel(R) Core(TM) i5-8250U CPU 1.80 GHz processor, Windows 10 64-bit operating system, × 64 based processor and 8 GB RAM."

"The proposed scheme is flexible in terms of dimension and size of message to be encrypted."

"Thorough testing was done on different parameter value to ensure the correctness of proposed scheme."

Security Analysis of AKAME Scheme

"The proposed scheme is robust against different security attacks."

"That there is no particular attack using the meticulous algebraic structure of involving rings, corresponding security parameters lambda are estimated using the lwe-estimator-9302d4204b4f by [77]. In this section we will discuss about possible attacks on lattice based schemes."

"The proposed scheme is secure from Chosen plaintext attack (CPA) [72]."

"Further, the Chosen ciphertext attack (CCA) secure version of proposed scheme can be obtained by using Fujisaki-Okamoto transform [72]."

"The scheme is 100% safe from decryption failure attacks since the hash digest matching ensure the successful key-agreement [88]."

"This is very important from security point of view as decryption failure causes many other attacks leading disclosure of secret key [89]."

"The decryption failure and man-in-the-middle attack is avoided by adding the concept of matching SHA hash digest of common secret key, b and b′ [90]."

Comparative Analysis of AKAME Scheme with Existing Scheme

"Comparative analysis of proposed scheme is done on the basis of key size and time taken for different operations."

"We compared our proposed scheme with the existing RLWE-based encryption and key exchange schemes."

"The message encryption and decryption time is lesser than the existing schemes."

Conclusions and Future Work

"This paper presents a novel quantum-secure authenticated key-agreement and message encryption scheme."

"The proposed non-interactive scheme is secure against both quantum and classical computers."

"The major advantage of AKAME scheme over existing classical cryptographic primitives are, (i) it is forward secure since data encrypted using this scheme remain robust in quantum era too."

"(ii) Reduces computational and communication overhead, (iii) the scheme is less complex as it make use of simple polynomial operations and non-prime modulus, (iv) AKAME is performance-wise efficient and secure as it uses discrete structure ring, (v) the transmitted messages over network are comparatively smaller than existing RLWE schemes, (vi) the proposed scheme is flexible in message size for encryption."

"The proposed AKAME scheme makes a successful key-agreement with high message throughput, zero failure rate and high security."

Review and Analysis of Classical Algorithms and Hash-Based Post-Quantum Algorithm [91]

This is a machine-generated summary of:

Noel, Moses Dogonyaro; Waziri, Victor Onomza; Abdulhamid, Shafii Muhammad; Ojeniyi, Joseph Adebayo: Review and analysis of classical algorithms and hash-based post-quantum algorithm [91].

Published in: Journal of Reliable Intelligent Environments (2021).

Link to original: https://doi.org/10.1007/s40860-021-00155-0

If you want to cite the papers, please refer to the original.

For technical reasons we could not place the page where the original quote is coming from.

Abstract-Summary

"Digital signature algorithms such as Rivest–Shamir–Adleman (RSA) and elliptic curve digital signature algorithm (ECDSA) are the commonly used algorithms to secure data in the public key infrastructure and other computing devices."

"The authors carried out the review analysis of two classical algorithms (RSA, ECDSA) and hash-based signature schemes; Winternitz one time signature (W-OTS) and Merkle signature (MSS), their security strength, efficiency in terms of key generation time, signature generation and verification time."

"The results showed that the two classical algorithms perform better in terms of the efficiency in key generation time, signature generation and verification time."

"The key generation time, signature generation and verification time increases when the key length increases."

"The security of the classical algorithms improved when the key length increase."

"The hash-based signature schemes in this research were considered to be the best alternative algorithms suitable for public key infrastructures considering the security properties exhibited by them."

Introduction

"This development aroused the interest of researchers in the field of post-quantum cryptography with emphasis mostly focusing on Hash-based signature schemes."

"The reason might be that the security property of hash-based signatures schemes are well known and can withstand any attack originating from a quantum computer."

"The aim of this research work is to carry out comparative analysis and review on some selected classical algorithms (Rivest–Shamir–Adleman and elliptic curve digital signature algorithm) with hash-based signature schemes (Winternitz one-time signature scheme and Merkle signature scheme) to ascertain their security properties, attack vectors, and efficiency to justify the need for the paradigm shift from classical algorithms to post-quantum hash-based signature schemes."

"Broad understanding of the security parameters of the chosen classical algorithms and the selected hash-based signature schemes."

"The limitations of Classical algorithms in the post-quantum era and thereby established the need for the paradigm shift from classical algorithm to the hash-based signature scheme."

Review of Related Literatures

"The work of [92] focused on the comparative analysis of post-quantum hash-based signature schemes."

"The significance of the study is to come up with a clear distinction on the security of hash-based signature scheme over asymmetric cryptographic algorithms such as RSA and ECDSA."

"The hash-based signature schemes where considered in this research as a good candidate in the post-quantum era because; their security properties are well understood, they do not rely on number-theoretic security or structured hardness assumptions, and are forward secure."

"The authors considered the review of two generic hash-based signature schemes to be compared with two classical schemes (RSA and ECDSA)."

"Hash-based Signature Schemes in recent time gained the attention of many researchers because they are considered the most suitable replacement for classical algorithms as discussed in [93]."

Methodology

"The first approach is the comparison that is based on the mathematical illustrations of the algorithms, attacks on these algorithms and their security properties."

"The efficiency (in terms of time taken to generate and verify signatures) of both the classical and hash-based signature algorithms would be implemented on a laptop computer."

"To decrypt the message M, the security of RSA is based on the hard computations involved in factoring very large primes [7], and increment in the key sizes (large key size) gives a strong security of data by the algorithm."

"The variable inputs used are key sizes for the RSA and ECDSA and then the output results which comprise of the key generation time, signature generation and verification time are recorded in seconds."

"The performance of the four algorithms (RSA, ECDSA, W-OTS, MSS) were implemented using Java programming language."

Discussion of Results

"Different key sizes were used and the time was recorded in seconds to generate and verify the signature."

"The aim is to obtain key generation time, signature generation and confirmation time measured in seconds for all the algorithms."

"When RSA key size was 15,360 KB, the key generation time was highest (approx."

"Although when RSA has its highest key size, the time taken to generate the signature was higher than ECDSA."

"The algorithms for key generation, signature generation and confirmation time would be calculated."

"The algorithms used in the Merkle Signature Scheme for key generation, signature generation and verification are illustrated here."

"The same is with the signature generation time of ECDSA and RSA which is faster than MSS."

"The development of a quantum algorithm by Peter Shor to solve the DLP and IFP of both RSA and ECDSA in polynomial time makes the two classical algorithms insecure."

Conclusion

"The rationale for conducting this research is to gain an understanding of the basic security strength of hash-based signature schemes over conventional algorithms."

"The study also revealed the efficiency of hash-based signature algorithms in terms of key generation time, signature generation and verification time."

"The study further gives an in-depth understanding of the limitations of conventional algorithms over the hash-based signatures schemes."

"Future research work is to consider other post-quantum signature schemes (such as lattice-based, code-based, and multivariate polynomial-based schemes), compare and evaluate these schemes with Hash-based signature schemes in terms of efficiency and security in blockchain technology and other constrain devices."

"The authors are currently working in reducing high latency in hash-based signature schemes for their implementation in light weight devices."

Breaking LWC Candidates: sESTATE and Elephant in Quantum Setting [94]

This is a machine-generated summary of:

Shi, Tairong; Wu, Wenling; Hu, Bin; Guan, Jie; Wang, Sengpeng: Breaking LWC candidates: sESTATE and Elephant in quantum setting [94].

Published in: Designs, Codes and Cryptography (2021).

Link to original: https://doi.org/10.1007/s10623-021-00875-7

If you want to cite the papers, please refer to the original.

For technical reasons we could not place the page where the original quote is coming from.

Abstract-Summary

"The Report on Lightweight Cryptography specifically asks that submissions be quantum safe when long-term security is needed."

"Most candidates, including sESTATE and Elephant, make no claim regarding security with respect to quantum attacks."

"We propose quantum key recovery attacks on those second-round candidates."

"The attack is based on the combination of a quantum extracting method and quantum square attack."

"For Elephant mode, which internally uses a permutation masked by linear feedback shift registers (LFSRs) similar to the masked Even-Mansour construction proposed in EUROCRYPT 2016, we launch the attack based on the quantum attack proposed by Bonnetain and others, which depends on Simon's algorithm without superposition queries and Grover's algorithm."

"Our attack is generic and independent of internal permutation; hence, we obtain the quantum attacks on all instances with a tradeoff of classical and quantum queries."

Introduction

"Bonnetain and others [95] proposed offline Simon's algorithm that allows applying some of the previous attacks in the Q1 model."

"We combined Simon's algorithm in the Q2 model and the quantum square attack on six-round AES [96]."

"The quantum attack of EM will not work because the nonce is always of fixed length m smaller than the scale of permutation n. To overcome it, we turn the structure into a variant of FX construction by dividing the masked whitening key into two parts, and obtain a valid Q2 attack based on [97], which made a notable combination of Simon's and Grover's algorithm to recover the whitening key of FX construction."

"We discuss several issues, including the use of block ciphers, the ways of masking, and length setting in AE mode, to avoid the corresponding quantum attacks and improve the performance of current schemes."

[Section 1]

"sESTATE uses two tweakable block ciphers: E and a round-reduced variant of E, represented by F. The tweakable block cipher F is used to process non-last blocks in the tag-generation phase."

"The tweakable block cipher E is used for all other cipher calls, i.e., to process the last block in tag generation and the full encryption phase."

"In the tag-generation phase, F is always employed with tweak value 15, and E with tweak values 2, 4, and 6 if the last block is full, or 3, 5 and 7 if the last block is partial."

"sESTATE_TweAES-128-6, the only recommended scheme for sESTATE mode, is obtained by TweAES-128 and a round-reduced version of TweAES-128, called a TweAES-128-6 tweakable block cipher."

"TweAES-128 is a tweakable variant of AES-128 block cipher with 4-bit tweak t and 128-bit key K, and is identical to AES-128 except for two AddTweak functions used at intervals of two rounds."

[Section 2]

"Before explaining the key-recovery attack on sESTATE_TweAES-128-6, we formally outline the high-level idea of the attack."

"We want to start the attack from TweAES-128-6."

"We face two problems: extracting TweAES-128-6 from the mode and efficiently recovering the key."

"For the first problem, we apply the quantum Simon's algorithm in the Q2 model to extract a TweAES-128-6 block cipher, which is impossible in a classical setting."

"For the second problem, we adopt the quantum square attack in the Q1 model on 6-round AES improved by a partial sum technique [96, 98]."

"We apply Simon's algorithm to extract the information of the inner part as a data provider, giving us partial access to TweAES-128-6."

"Algorithm 3 outlines the classical square attack of sESTATE_TweAES-128-6."

"In the extracting phase, we perform Q2 queries to sESTATE approximately times to provide enough input-output pairs for the quantum square attack."

Quantum Key-Recovery Attack on Elephant

"Our attack model focuses on the encryption part in Elephant mode, and we describe only that."

"We briefly outline the key-recovery attack on Elephant with the technique in [97]."

"Key-recovery attack on Elephant in the Q2 model with reduced queries."

"We give the details of the tradeoff for Q1 attack on Elephant."

"Hope that the total complexity can be lower than that of Grover's generic attack on key."

"The effective key lengths of the recommended schemes are not up to the quantum security margin in the Q2 model."

Discussion

"On sESTATE mode, we showed how to extract the potential weak part using a quantum algorithm, which can sometimes be extended to a full key-recovery attack."

"We raise the following issues in the hope of providing a reference or reinforcing the design of lightweight AE, particularly when considering quantum attacks."

"Use the round-reduced (tweakable) block cipher with caution. It is the prevalent way to instantiate the block cipher-based AE mode with reasonable block ciphers."

"To optimize an efficient (tweakable) block cipher design for AE is another important step."

"We have many examples, such as the seminal work of Grover's algorithm reducing the effective key-length of any cryptographic cipher by a factor of two, the recent results of breaking the EM and FX constructions with whitening keys, and our work on Elephant with masked EM construction."

A Novel Digital Contents Privacy Scheme Based on Quantum Harmonic Oscillator and Schrodinger Paradox [99]

This is a machine-generated summary of:

Alghafis, Abdullah; Waseem, Hafiz Muhammad; Khan, Majid; Jamal, Sajjad Shaukat; Amin, Muhammad; Batool, Syeda Iram: A novel digital contents privacy scheme based on quantum harmonic oscillator and schrodinger paradox [99].

Published in: Wireless Networks (2020).

Link to original: https://doi.org/10.1007/s11276-020-02363-7

If you want to cite the papers, please refer to the original.

For technical reasons we could not place the page where the original quote is coming from.

Abstract-Summary

"The security of digital information is one of the most unavoidable issue in current digital world."

"The existing world facing number of digital information breaches due to weak utilization of confidentiality preserving techniques."

"We have designed a new mechanism for the security of secret digital information for instance digital images."

Introduction

"The advancement in digital world can make it possible to transmit an information from one place to another with ease never before with such advancement in science and technology."

"Security of these digital information is also an inevitable concern."

"Before 1950′s different confidentiality schemes were utilized in order to secure information."

"From last two decades several digital information confidentiality mechanisms were designed in order to provide the security to digital secrets of any nations."

"Cybersecurity manages the security of computer systems from attacks that could bargain with the hardware equipment, software and digital information."

"Quantum digital security is the new emerging area of science that reviews all the perspectives affecting the security and, protection of communication and computations instigated by the development of quantum innovations [100–104]."

"We have designed a novel digital content confidentiality scheme by utilizing the quantum harmonic oscillations."

Basic Preliminaries

"The fundamental concepts of quantum harmonic oscillations and Schrodinger paradox demonstrated in subsections of 2."

"At adequately small energies, the harmonic oscillator as represented by the laws of quantum mechanics, referred as the quantum harmonic oscillator, differs significantly from its depiction as indicated by the laws of classical physics [105, 106]."

"As one of the few significant quantum mechanical frameworks whose elements can be determined exactly, the quantum harmonic oscillator as often as possible fills in as a basis for describing many real-world implications, such as vibration control, random number generation etc., [107]."

"Physicists have contrived a variety of the notable Schrodinger's cat psychological test that involves several players who comprehend quantum hypothesis."

"The quantum harmonic oscillator is one of the foundation problems of quantum mechanics."

"The digital contents followed by the 8-bit sequence data of quantum harmonic oscillator with a specified energy state produces 8-bit cipher sequence."

Performance and Security Evaluations for Anticipated Structure

"To assert the relationship among adjoining pixels in plain and enciphered image, 10,000 arrangements of two close-by pixels from each content initially selected [108, 109]."

"The relationship among various couples of plain and encoded images evaluated by computing the two-dimensional correlation coefficients among the plain and enciphered images [110, 111]."

"We have examined the structural similarity index (SSIM), normalized cross-correlation (NCC) and structural contents (SC) between plain and enciphered images in order to evaluate the structure distinction among various digital contents from reference."

"The average absolute difference among plain and enciphered contents determined here by evaluating the MAE, and its esteem must be more prominent to improve the encryption quality."

"Both MSE and PSNR used to relate the image encryption quality, while MSE indicates the aggregate square error measure and PSNR exhibits the peak error estimation between the plain and enciphered contents."

Forensics Analyses

"To perform the linear assaults, cryptanalyst needs to perceive the immediate association between specific bits of the plain image, enciphered image and a key."

"To certify the strength against differential assaults for the foreseen plan, an alteration of one pixel in the plain image alters the encoded image for equating with a probability of half-pixel modifying."

"For a specific objective to evaluate the effect of minor alteration in the plain image on its encode one, the number of pixels changing rate (NPCR) bound together to locate the unified average change intensity (UACI) [112, 113]."

"As the intensity of noise varies from 0.000001 to 0.000007, the PSNR assessment has very minute change or we can say its roughly equivalent to the original which depicts the proposed structure has incredible strength against noise assaults."

Concluding Remarks and Future Prospects

"The ability to communicate securely and process adequately is more significant than ever to society."

"Over the next 5–10 years, we will see a transition of new potential outcomes, as quantum advancements become part of this mainstream processing and communicating landscape."

"The acknowledgment of such complex frameworks of classical style and quantum correspondence must rely upon a solid novel foundation that can predict and deal with the complexities of real time implementation and novel applications."

"The proposed plan is reasonable for real time applications because of small processing time and better ability to hostile the assaults and appropriate execution than other encryption frameworks."

References

1. Hekkala J, Muurman M, Halunen K, Vallivaara V (2023) Implementing Post-Quantum Cryptography for Developers. SN Computer Science. https://doi.org/10.1007/s42979-023-01724-1
2. https://www.openssl.org/roadmap.html. Accessed 5 Aug 2022
3. Bos J, Costello C, Ducas L, Mironov I, Naehrig M, Nikolaenko V, Raghunathan A, Stebila D (2016) Frodo: take off the ring! practical, quantum-secure key exchange from lwe. In: Proceedings of the 2016 ACM SIGSAC Conference on Computer and Communications Security, pp 1006–1018
4. https://csrc.nist.gov/Projects/post-quantum-cryptography/post-quantum-cryptography-standardization/Call-for-Proposals. Accessed 8 Aug 2022
5. Tan TG, Szalachowski P, Zhou J (2022) Challenges of post-quantum digital signing in real-world applications: a survey. Int J Inf Secur 21:937–952. https://doi.org/10.1007/s10207-022-00587-6
6. Proos J, Zalka C (2003) Shor's discrete logarithm quantum algorithm for elliptic curves. arXiv preprint arXiv:quant-ph/0301141
7. Shor PW (1999) Polynomial-time algorithms for prime factorization and discrete logarithms on a quantum computer. SIAM Rev 41:303–332
8. NIST: Submission requirements and evaluation criteria for the post-quantum cryptography standardization process. Online: https://csrc.nist.gov/CSRC/media/Projects/Post-Quantum-Cryptography/documents/call-for-proposals-final-dec-2016.pdf. Accessed Apr 2021 (2016)
9. Cooper DA, Apon DC, Dang QH, Davidson MS, Dworkin MJ, Miller CA (2020) Recommendation for stateful hash-based signature schemes. NIST Spec Publ 800:208
10. Manulis M, Bridges CP, Harrison R, Sekar V, Davis A (2021) Cyber security in new space. Int J Inf Secur 20(3):287–311
11. Tan TG, Zhou J (2021) Layering quantum-resistance into classical digital signature algorithms. In: International conference on information security. Springer, pp 26–41

12. Banerjee U, Ukyab TS, Chandrakasan AP (2019) Sapphire: a configurable crypto-processor for post-quantum lattice-based protocols. arXiv preprint arXiv:1910.07557
13. Dang VB, Farahmand F, Andrzejczak M, Mohajerani K, Nguyen DT, Gaj K (2020) Implementation and benchmarking of round 2 candidates in the nist post-quantum cryptography standardization process using hardware and software/hardware co-design approaches. Cryptology ePrint Archive: Report 2020/795
14. Hülsing A, Rijneveld J, Schwabe P (2016) Armed sphincs. In: Public-Key Cryptography–PKC 2016. Springer, Berlin/Heidelberg, pp 446–470
15. multiple: Post-quantum crypto library for the arm cortex-m4. Online: https://github.com/mupq/pqm4. Accessed Apr 2021 (2020)
16. Oder T, Pöppelmann T, Güneysu T (2014) Beyond ecdsa and rsa: lattice-based digital signatures on constrained devices. In: 2014 51st ACM/EDAC/IEEE Design Automation Conference (DAC). IEEE, pp 1–6
17. Albrecht MR, Hanser C, Hoeller A, Pöppelmann T, Virdia F, Wallner A (2019) Implementing rlwe-based schemes using an rsa co-processor. IACR Transactions on Cryptographic Hardware and Embedded Systems:169–208
18. Boorghany A, Sarmadi SB, Jalili R (2015) On constrained implementation of lattice-based cryptographic primitives and schemes on smart cards. ACM Trans Embed Comput Syst (TECS) 14(3):1–25
19. Ahilan A, Jeyam A (2022) Breaking barriers in conventional cryptography by integrating with quantum key distribution. Wirel Pers Commun 129:549–567. https://doi.org/10.1007/s11277-022-10110-8
20. da Silva TF, Xavier GB, Temporão GP, von der Weid JP (2012) Real-time monitoring of single-photon detectors against eavesdropping in quantum key distribution systems. Opt Express 20(17):18911–18924
21. Moizuddin M, Winston J, Qayyum M (2017) A comprehensive survey: quantum cryptography. In: 2017 2nd International Conference on Anti-Cyber Crimes (ICACC). IEEE, pp 98–102. https://doi.org/10.1109/Anti-Cybercrime.2017.7905271
22. Rawal S, Sahadeo P (2022) A quantum resistant anonymous poxy signature scheme. Sādhanā 47. https://doi.org/10.1007/s12046-022-01808-3
23. Jiang Y, Kong F, Ju X (2010) Lattice-based proxy signature. In: Proceeding of Computational Intelligence and Security (CIS), pp 382–385
24. Wang C, Qi M (2011) Lattice-based proxy signature scheme. J Inf Comput Sci 8(12):2451–2458
25. Xia F, Yang B, Ma S (2011) Lattice-based proxy signature scheme. J f Hunan Univ (Nat Sci) 38(6):84–88
26. Zhang L, Sang Y (2012) A lattice-based identity-based proxy signature from bonsai trees. Int J Adv Comput Technol 4(20):99–104
27. Kim KS, Hong D, Jeong IR (2013) Identity-based proxy signature from lattices. Commun Netw 15(1):17–17
28. Wu F, Wang Y, Zhang X, Wang W, Zheng Z (2018) Identity-based proxy signature over NTRU lattice. Int J Commun Syst 32(3)
29. Giron AA, Custódio R, Rodríguez-Henríquez F (2022) Post-quantum hybrid key exchange: a systematic mapping study. J Cryptogr Eng 13:71–88. https://doi.org/10.1007/s13389-022-00288-9
30. Schanck JM, Whyte W, Zhang Z (2016) Circuit-extension handshakes for tor achieving forward secrecy in a quantum world. Proc Priv Enhancing Technol 2016(4):219–236
31. Barker E, Chen L, Davis R (2020) Recommendation for key-derivation methods in key-establishment schemes (draft). NIST Spec Publ 800:56C
32. Datta A, Derek A, Mitchell JC, Warinschi B (2006) Key exchange protocols: security definition, proof method and applications. IACR Cryptol ePrint Arch 2006:56
33. Paquin C, Stebila D, Tamvada G (2020) Benchmarking post-quantum cryptography in tls. In: Ding J, Tillich J-P (eds) Post-quantum cryptography. Springer, Cham, pp 72–91

34. Bindel N, Brendel J, Fischlin M, Goncalves B, Stebila D (2019) Hybrid key encapsulation mechanisms and authenticated key exchange. In: Ding J, Steinwandt R (eds) Post-quantum cryptography. Springer, Cham, pp 206–226
35. Giacon F, Heuer F, Poettering B (2018) Kem combiners. In: Abdalla M, Dahab R (eds) Public-Key Cryptography—PKC 2018. Springer, Cham, pp 190–218
36. Ghosh S, Kate A (2015) Post-quantum forward-secure onion routing. In: Malkin T, Kolesnikov V, Lewko AB, Polychronakis M (eds) Applied cryptography and network security. Springer, Cham, pp 263–286
37. Crockett E, Paquin CS (2019) Prototyping post-quantum and hybrid key exchange and authentication in tls and ssh. Cryptology ePrint Archive, Report 2019/858
38. Schwabe P, Stebila D, Wiggers T (2020) Post-quantum tls without handshake signatures. IACR Cryptol ePrint Arch 2020:534
39. Stebila D, Mosca M (2017) Post-quantum key exchange for the internet and the open quantum safe project. In: Avanzi R, Heys H (eds) Selected Areas in Cryptography—SAC 2016. Springer, Cham, pp 14–37
40. Pirnay, Niklas; Pappa, Anna; Seifert, Jean-Pierre learning classical readout quantum PUFs based on single-qubit gates. Quantum Mach Intell (2022). doi:https://doi.org/10.1007/s42484-022-00073-1, 4
41. Phalak K, Saki AA, Alam M, Topaloglu RO, Ghosh S (2021) Quantum puf for security and trust in quantum computing. IEEE J Emerg Sel Top Circuits Syst 11:333–342. https://doi.org/10.1109/JETCAS.2021.3077024
42. Doosti M, Kumar N, Delavar M, Kashefi E (2021) Client-server identification protocols with quantum puf. ACM Trans Quantum Comput 2:1–40. https://doi.org/10.1145/3484197
43. Hinsche M, Ioannou M, Nietner A, Haferkamp J, Quek Y, Hangleiter D, Seifert JP, Eisert J, Sweke R (2021) Learnability of the output distributions of local quantum circuits. arXiv preprint arXiv:2110.05517
44. Gollakota A, Liang D (2021) On the hardness of pac-learning stabilizer states with noise. arXiv preprint:arXiv:2102.05174
45. Kearns M (1998) Efficient noise-tolerant learning from statistical queries. J ACM 45:983–1006. https://doi.org/10.1145/293347.293351
46. Škorić B (2012) Quantum readout of physical unclonable functions. Int J Quantum Inf 10:1250001. https://doi.org/10.1142/S0219749912500013
47. Ghosh S, Watrous J (2022) Complexity limitations on one-turn quantum refereed games. Theory Comput Syst 67:383–412. https://doi.org/10.1007/s00224-022-10105-9
48. Fortnow L, Impagliazzo R, Kabanets V, Umans C (2008) On the complexity of succinct zero-sum games. Comput Complex 17:353–376
49. Gutoski G, Wu X (2013) Parallel approximation of min-max problems. Comput Complex 2(22):385–428
50. Althöfer I (1994) On sparse approximations to randomized strategies and convex combinations. Linear Algebra Appl 199:339–355
51. Lipton R, Young N (1994) Simple strategies for large zero-sum games with applications to complexity theory. In: Proceedings of the Twenty-Sixth Annual ACM Symposium on Theory of Computing, pp 734–740
52. Al-saggaf AA, Sheltami T, Alkhzaimi H, Ahmed G (2022) Lightweight two-factor-based user authentication protocol for IoT-enabled healthcare ecosystem in quantum computing. Arab J Sci Eng 48:2347–2357. https://doi.org/10.1007/s13369-022-07235-0
53. Yang W, Wang S, Zheng G, Yang J, Valli C (2019) A privacy-preserving lightweight biometric system for internet of things security. IEEE Commun Mag. https://doi.org/10.1109/MCOM.2019.1800378
54. Ayub MF, Mahmood K, Kumari S, Sangaiah AK (2020) Lightweight authentication protocol for e-health clouds in IoT based applications through 5G technology. Digit Commun Netw. https://doi.org/10.1016/j.dcan.2020.06.003

55. Rehman HU, Ghani A, Chaudhry SA et al (2021) A secure and improved multi server authentication protocol using fuzzy commitment. Multimed Tools Appl 80:16907–16931. https://doi.org/10.1007/s11042-020-09078-z
56. Mohammed AJ, Yassin AA (2019) Efficient and flexible multi-factor authentication protocol based on fuzzy extractor of administrator's fingerprint and smart mobile device. Cryptography. https://doi.org/10.3390/cryptography3030024
57. Taher BH, Jiang S, Yassin AA, Lu H (2019) Low-overhead remote user authentication protocol for IoT based on a fuzzy extractor and feature extraction. IEEE Access 7:256. https://doi.org/10.1109/ACCESS.2019.2946400
58. Sahu AK, Sharma S, Puthal D (2020) Lightweight multi-party authentication and key-agreement protocol in IoT based e-healthcare service. ACM Trans Multimed Comput Commun Appl 17:1–20. https://doi.org/10.1145/3398039
59. Gupta DS, Islam SH, Obaidat MS, Karati A, Sadoun B (2020) LAAC: lightweight lattice-based authentication and access control protocol for E-health systems in IoT environments. IEEE Syst J 15:3620–3627. https://doi.org/10.1109/jsyst.2020.3016065
60. Regev O (2009) On lattices, learning with errors, random linear codes, and cryptography. J ACM (JACM) 56(6):34–40
61. Al-Saggaf AA (2021) A post-quantum fuzzy commitment scheme for biometric template protection: an experimental study. IEEE Access 9:110952–110961. https://doi.org/10.1109/ACCESS.2021.3100981
62. Mukherjee S, Gupta DS, Biswas GP (2019) An efficient and batch verifiable conditional privacy-preserving authentication scheme for VANETs using lattice. Computing 101:1763–1788. https://doi.org/10.1007/s00607-018-0689-3
63. Mittal G, Kumar S, Kumar S (2022) A quantum secure ID-based cryptographic encryption based on group rings. Sādhanā 47. https://doi.org/10.1007/s12046-022-01806-5
64. Milies C, Sehgal S (2002) An introduction to group rings. Springer, Dordrecht
65. Boneh D, Franklin M (2003) Identity based encryption from the Weil pairing. SIAM J Comput 32:586–615
66. Hoffstein J, Pipher J, Silverman J (2008) An introduction of mathematical cryptography. Springer, New York
67. Mittal G, Kumar S, Narain S, Kumar S (2021) Group rings based public key cryptosystems. J Discret Math Sci Cryptogr Online First 25:1683–1704. https://doi.org/10.1080/09720529.2020.1796868
68. Miller V (1986) Short programs for functions on curves. http://crypto.stanford.edu/miller/miller.pdf
69. Hurley B, Hurley T (2011) Group ring cryptography. Int J Pure Appl Math 69:67–86
70. Choudhary S, Gupta A (2022) AKAME: a post-quantum authenticated key-agreement and message encryption scheme based on ring-LWE. Int J Inf Technol 14:1669–1676. https://doi.org/10.1007/s41870-022-00888-y
71. Shor PW (1994) Algorithms for quantum computation: discrete logarithms and factoring. In: Foundations of computer science, 1994, proceedings, 35th annual symposium on. IEEE, pp 124–134
72. Peikert C (2015) A decade of lattice cryptography, Cryptology ePrint Archive, Report 2015/939
73. NIST (2016) Post-quantum crypto project. http://csrc.nist.gov/groups/ST/post-quantum-crypto/
74. NIST (2016) Submission requirements and evaluation criteria for the post-quantum cryptography standardization process, Post-Quantum Crypto Standardization, Call For Proposals Announcement
75. Sendhil R, Amuthan A (2021) Contextual fully homomorphic encryption schemes-based privacy preserving framework for securing fog-assisted healthcare data exchanging applications. Int J Inf Tecnol 13:1545–1553. https://doi.org/10.1007/s41870-021-00704-z

76. Albrecht MR (2017) On dual lattice attacks against small-secret LWE and parameter choices in HElib and SEAL. In: EUROCRYPT. Springer, pp 103–129
77. Albrecht MR, Ducas L, Herold G, Kirshanova E, Postlethwaite E, Stevens M (2019) The general sieve kernel and new records in lattice reduction, Cryptology ePrint Archive, Report 2019/089, In Eurocrypt
78. Regev O (2010) The learning with errors problem (invited survey). In CCC, pp 191–204
79. Albrecht MR, Player R, Scott S (2015) On the concrete hardness of learning with errors. J Math Cryptol 9(3):169–203
80. Lyubashevsky V, Peikert C, Regev O (2010) On ideal lattices and learning with errors over rings. In: EUROCRYPT. Springer, Berlin/Heidelberg, pp 1–23
81. Fluhrer S (2016) Cryptanalysis of ring-LWE based key exchange with key share reuse, Cryptology ePrint Archive, Report 2016/085
82. Micciancio D, Walter M (2016) Practical, predictable lattice basis reduction. In: EUROCRYPT. Springer, Berlin/Heidelberg, pp 820–849
83. Micciancio D (2010) Duality in lattice cryptography, In Public key cryptography
84. Hoffstein J, Pipher J, Silverman J (1998) NTRU: a ring-based public key cryptosystem. Algorithmic Number Theo 1998:267–288
85. Ding J, Xie X, Lin X (2012) A simple provably secure key exchange scheme based on the learning with errors problem. IACR EPrint Archive 688
86. Alkim E, Ducas L, Pöppelmann T, Schwabe P (2016) Postquantum key exchange—a new hope. In USENIX Security Symposium, pp 327–343
87. Choudhary S, Gupta A (2018) Efficient implementation and analysis of ring-LWE quantum-secure key exchange protocol. Int J Adv Stud Sci Res 3(9) Available at SSRN: https://ssrn.com/abstract=3315427
88. Toom AL (1963) The complexity of a scheme of functional elements realizing the multiplication of integers. In Soviet Math Doklady 3:714–716
89. Dnvers JP, Guo Q, Johansson T, Nilsson A, Vercauteren F, Verbauwhede I (2019) On the impact of decryption failures on the security of LWE/LWR based schemes, In PKC: 565–589
90. Choudhary S, Gupta A (2022) HybridPKE: a forward-secure non-interactive quantum-safe hybrid key exchange scheme. Eng Sci Technol Int J 34:101094., ISSN 2215-0986. https://doi.org/10.1016/j.jestch.2022.101094
91. Noel MD, Waziri VO, Abdulhamid SM, Ojeniyi JA (2021) Review and analysis of classical algorithms and hash-based post-quantum algorithm. J Reliab Intell Environ 8:397–414. https://doi.org/10.1007/s40860-021-00155-0
92. Katz J (2016) Analysis of a proposed hash-based signature standard. International conference on research in security standardisation, vol 10074. Springer, Cham, pp 261–273
93. Suhail S, Hussain R, Khan A, Hong CS (2020) On the role of hash-based signatures in quantum-safe internet of things: current solutions and future directions. IEEE Internet Things J 8(1):1–7. https://doi.org/10.1109/JIOT.2020.3013019
94. Shi T, Wu W, Hu B, Guan J, Wang S (2021) Breaking LWC candidates: sESTATE and elephant in quantum setting. Des Codes Crypt 89:1405–1432. https://doi.org/10.1007/s10623-021-00875-7
95. Bonnetain X, Hosoyamada A, Naya-Plasencia M, Sasaki Y, Schrottenloher A (2019) Quantum attacks without superposition queries: the offline Simon's algorithm. In: Advances in cryptology-ASIACRYPT 2019-25th International Conference on the Theory and Application of Cryptology and Information Security, Kobe, Japan, December 8–12, 2019, Proceedings, Part I, pp 552–583. https://doi.org/10.1007/978-3-030-34578-5_20
96. Bonnetain X, Naya-Plasencia M, Schrottenloher A (2019) Quantum security analysis of AES. IACR Trans Symmetric Cryptol 2019(2):55–93. https://doi.org/10.13154/tosc.v2019.i2.55-93
97. Leander G, May A (2017) Grover meets Simon—quantumly attacking the fx-construction. In: Advances in Cryptology-ASIACRYPT 2017-23rd International Conference on the Theory and Applications of Cryptology and Information Security, Hong Kong, China, December 3–7, 2017, Proceedings, Part II, pp. 161–178. https://doi.org/10.1007/978-3-319-70697-9_6

98. Ferguson N, Kelsey J, Lucks S, Schneier B, Stay M, Wagner DA, Whiting D (2000) Improved cryptanalysis of rijndael. In: Fast Software Encryption, 7th International Workshop, FSE 2000, New York, NY, USA, April 10–12, 2000, Proceedings, pp. 213–230. https://doi.org/10.1007/3-540-44706-7_15
99. Alghafis A, Waseem HM, Khan M, Jamal SS, Amin M, Batool SI (2020) A novel digital contents privacy scheme based on quantum harmonic oscillator and schrodinger paradox. Wirel Netw 30:6217–6236. https://doi.org/10.1007/s11276-020-02363-7
100. Boneh D, Zhandry M (2013) Secure signatures and chosen ciphertext security in a quantum computing world. In: Annual cryptology conference. Springer, Berlin, pp 361–379
101. Gagliardoni T, Hülsing A, Schaffner C (2016) Semantic security and indistinguishability in the quantum world. In: Annual International Cryptology Conference. Springer, Berlin, pp 60–89
102. Kaplan M, Leurent G, Leverrier A, Naya-Plasencia M (2016) Breaking symmetric cryptosystems using quantum period finding. In: Annual International Cryptology Conference. Springer, Berlin, pp 207–237
103. Lydersen L, Wiechers C, Wittmann C, Elser D, Skaar J, Makarov V (2010) Hacking commercial quantum cryptography systems by tailored bright illumination. Nat Photonics 4(10):686–689
104. Xu F, Arrazola JM, Wei K, Wang W, Palacios-Avila P, Feng C et al (2015) Experimental quantum fingerprinting with weak coherent pulses. Nat Commun 6:8735
105. Griffiths DJ, Schroeter DF (2018) Introduction to Quantum Mechanics. Cambridge University Press, Cambridge
106. McQuarrie DA (2000) Statistical Mechanics. University Science Books, Sausalito CA
107. Shankar R (2012) Principles of Quantum Mechanics. Springer, Berlin
108. Khan M, Waseem HM (2019) A novel digital contents privacy scheme based on Kramer's arbitrary spin. Int J Theor Phys 58(8):2720–2743
109. Batool SI, Waseem HM (2019) A novel image encryption scheme based on Arnold scrambling and Lucas series. Multimed Tools Appl 78(19):27611–27637
110. Munir N, Khan M (2018) A generalization of algebraic expression for nonlinear component of symmetric key algorithms of any characteristic. In: 2018 International Conference on Applied and Engineering Mathematics (ICAEM). IEEE, pp 48–52
111. Rafiq A, Khan M (2018) Construction of new S-boxes based on triangle groups and its applications in copyright protection. Multimed Tools Appl 78(11):15527–15544
112. Khan M, Shah T (2014) A construction of novel chaos base nonlinear component of block cipher. Nonlinear Dyn 76(1):377–382
113. Khan M, Shah T, Batool SI (2016) A new implementation of chaotic S-boxes in CAPTCHA. SIViP 10(2):293–300

Chapter 4
Simulation of Quantum and Artificial Systems

Introduction by the Editor

Let us begin with a challenge. How to integrate or combine Artificial Intelligence and Quantum systems? This chapter describes quantum pattern recognition on real quantum processing units, quantum simulations of convolutional neural networks, genetic algorithms for finding bipolar single flux-quantum pulse sequences for qubit control, multi-quit states modular superconducting quantum processors and quantum natural language processors. Also, it introduces neural error mitigation which uses neural networks to improve estimates of ground states and ground-state observables from near-term quantum simulations. Additionally, deep rural networks with quantum layers are discussed. This chapter demonstrates the applications of the simulation framework in aiding the development of quantum algorithms and quantum error correction.

This chapter investigates swap test potential for pattern recognition tasks in Quantum Computing (QC). As of now, QC refers to the use of fundamental principles of physics such as superposition and entanglement to process information. The swap test is a basic algorithm used for comparing different or multiple states in Quantum Mechanics (QM). Using it in pattern recognition tasks would be an interesting idea especially because it can efficiently detect similarity among various objects. It is demonstrated that these tests were done using IBMQ noisy intermediate-scale quantum (NISQ) devices. Beyond this point Amplitude Transformed Quantum Convolutional Neural Network (ATQCNN) comes into play by addressing some limitations imposed purely by QCNN, particularly slow convergence and low training efficiency. Through experiments, ATQCNN does convergence faster on quantum cluster states. Then we perform entanglement concentration protocols to partially entangled subsets of highly entangled multi-qubit states, such as Z-states and cluster states (family of quantum states), which serve as valuable resources in measurement-based computation models. Utilizing Z-states and cluster states as

V. Natarajan (ed.), *Quantum Artificial Intelligence*,
https://doi.org/10.1007/978-981-96-5051-4_4

target states demonstrates the practical relevance of your work, as these states are known to be powerful resources in quantum information processing. The application of entanglement concentration protocols to partially entangled subsets of highly entangled multi-qubit states, specifically focusing on Z-states and cluster states, are known for their significance in measurement-based computation models. Leveraging highly entangled multi-qubit states as a resource in quantum computing is a strategic approach, as these states offer unique advantages for various quantum information processing tasks. Further, we are deep into neural error mitigation, which leverages neural networks to enhance the accuracy of ground state estimates and ground-state observables obtained from near-term quantum simulations. Then we discuss the challenge of scaling in superconducting quantum computing by proposing a modular architecture with low-loss interconnects.

In this chapter, we explore the intersection between Natural Language Processing (NLP) and Quantum Computing (QC). The advantage of Quantum Natural Language Processing (QNLP) is performance wise faster than the classical NLP. One of the purposes of this section is to discuss a model called Categorical Distributional Compositional (DisCoCat) as a basis for NLP framework which could be unknown to some readers. The DisCoCat framework performs well on text classification tasks. We also show how its mathematical background makes it possible for quantum instantiations. Today's world, QNLP plays an important role in text mining, automated dialogue, language translation, bioinformatics, text-to-voice and voice-to-text. Thereby, QNLP supports Large Language Models (LLMs) while handling vast amount of data.

For the first time ever performing an NLP task on a Noisy Intermediate-Scale Quantum (NISQ) processor using the DisCoCat mathematical framework is considered to be a significant achievement. This is an application of Category theory in NISQ.

We introduce several parameterized circuit ansätze which are inspired by mixing and phase separation in Quantum Alternating Operator Ansatz (QAOA). These ansätze are designed specifically keeping in mind compilation considerations for near-term superconducting quantum processors. Also provide context about QAOA and its significance in quantum multi-dimensional optimization problems. Although we discuss challenges or risks faced while implementing QAOA on near-term quantum processors particularly circuit depth and compilation considerations in this chapter.

There is an important concept in Artificial Intelligence called deep neural networks (DNNs), that aligns with Quantum (Q). We give context about deep learning and its successes across different fields. Are there advantages of QDNNs over DNNs? This book clarifies it. Of course, we have a motive for exploring quantum extensions of classical DNNs highlighting the potential advantages of quantum computing in enhancing representation power and computational efficiency. Further we describe the architecture of QDNN, emphasizing its use of quantum structured layers.

In this chapter, we have focused on the tensor network. It is a well-established framework for presenting multilinear functions over multidimensional arrays called tensors. It is widely applied in quantum physics, machine learning, and quantum computing. The tensor network has a concept termed as multi hypergraph which helps each node of a tensor network connected to multiple tensor nodes. Every tensor network is engaged with multiple tasks called the tensor network contraction. How to do contracting tensor networks? A general method is done through sequential pairwise contraction. That means, any two tensors from the tensor network are chosen and mixed or merged. This process reduces the number of tensor nodes in the tensor network by one. Repeatedly, while doing the same process for other tensor nodes. Finally, we end up at the one tensor node in the tensor network. This kind of approach reduces the time complexity by means of floating-point operations (FLOPS) and index slicing.

We present an algorithmic framework for efficiently contracting tensor networks, enabling the classical simulation of quantum computations previously considered infeasible. We highlight the challenges or open problems associated with contracting large tensor networks and the motivation for developing efficient algorithms to overcome these challenges.

In this chapter, let us present a numerical demonstration or case study showcasing the coherent interaction between quantum emitters and a sub-wavelength plasmonic nano-cavity. Our study reveals that the interplay between the near field of quantum emitters and the gap plasmon field induces a remarkable canalization of the emission. In particular, a cascade of metal-insulator-metal (MIM) nanostructures has been examined and interlocked through a nano-cavity containing the quantum emitters. Through numerical simulations or combined estimation, we observe a selective reshaping of the radiated field, highlighting the potential for controlling and manipulating quantum interactions at the nanoscale.

At the end of Chap. 4, we propose a scheme or method for processing quantum data that is applicable to improving the performance of quantum computing devices as they advance towards fault-tolerant universal quantum computers. Although the research work involves the discrimination and filtering of quantum states provided as input to the device, with potential applications in various quantum information technologies. To demonstrate the feasibility of fault-tolerant models, we perform a proof-of-principle implementation using a superconducting quantum processor available on the IBM Quantum Experience platform. Similarly many corporates and research industries have started designing Quantum platforms at a high range.

Machine Generated Summaries

Disclaimer: The summaries in this chapter were generated from Springer Nature publications using extractive AI auto-summarization: An extraction-based summarizer aims to identify the most important sentences of a text using an algorithm and uses those original sentences to create the auto-summary (unlike generative AI). As the constituted sentences are machine selected, they may not fully reflect the body of the work, so we strongly advise that the original content is read and cited. The auto generated summaries were curated by the editor to meet Springer Nature publication standards. To cite this content, please refer to the original papers.

Machine generated keywords: superconducting qubits, neural, entangle, fidelity, pattern, superconducting quantum, flux, tensor network, neural network, concentration, nisq, machine, interaction, pqcs, image.

Quantum Pattern Recognition on Real Quantum Processing Units [1]

This is a machine-generated summary of:

Das, Sreetama; Zhang, Jingfu; Martina, Stefano; Suter, Dieter; Caruso, Filippo: Quantum pattern recognition on real quantum processing units [1].

Published in: Quantum Machine Intelligence (2023).

Link to original: https://doi.org/10.1007/s42484-022-00093-x

If you want to cite the papers, please refer to the original.

For technical reasons we could not place the page where the original quote is coming from.

Abstract-Summary

"We investigate the possibility of realizing a quantum pattern recognition protocol based on swap test, and use the IBMQ noisy intermediate-scale quantum (NISQ) devices to verify the idea."

"We find that with a two-qubit protocol, swap test can efficiently detect the similarity between two patterns with good fidelity, though for three or more qubits, the noise in the real devices becomes detrimental."

"To mitigate this noise effect, we resort to destructive swap test, which shows an improved performance for three-qubit states."

"Due to limited cloud access to larger IBMQ processors, we take a segment-wise approach to apply the destructive swap test on higher dimensional images."

"We also present an experimental set up for applying destructive swap test using the nitrogen vacancy (NVs) center in diamond."

"We propose a protocol inspired from quantum associative memory, which works in an analogous way to supervised learning for performing quantum pattern recognition using destructive swap test."

Introduction

"With the advancement in quantum technologies, a growing interest has been directed towards implementing quantum systems for improved image processing."

"Some quantum image processing protocols have been proposed and tested, which show polynomial and exponential speed up; an example is quantum edge detection (Zhang and others [2]; Yao and others [3]; Cavalieri and Maio [4]; Xu and others [5])."

"Except these, quantum pattern recognition protocols based on the framework of classical Hopfield neural network (Neigovzen and others [6]), the hidden shift problem (Montanaro [7]), pixel gradient calculation (Zhang and others [8]) and Grover's algorithm (Jiang and others [9]; Soni and Rasool [10]; Tezuka and others [11]) have been proposed."

"When tested on real quantum systems, the protocols are inevitably subject to noise, which degrades their efficiency."

"We consider a quantum pattern recognition algorithm based on swap test."

"The swap test is used to calculate the closeness between two quantum states."

"In quantum image processing, the classical images are encoded in quantum states; hence, the swap test emerges as a very plausible approach to find similarity between two quantum images."

Encoding an Image in a Quantum State

"The first task in quantum image processing is to encode the pixel positions and corresponding pixel values of a classical image in a quantum state."

"We work with a 2D grayscale image of $N = P \times Q$ pixels, which is classically encoded in a $P \times Q$ dimensional matrix M, the matrix element M(i,j) denoting the pixel value at position (i,j)."

"The images can also be grayscale, where the varying pixel values denote different shades of gray."

"In a binary image, all the pixel values are 0 or 1."

"The conversion from a grayscale to a binary image can be achieved by choosing a suitable threshold pixel value p in between the minimum and maximum pixel values of the greyscale image."

"In classical image processing, a brute-force method is to measure the pixel values of all the N pixels."

Swap Test

"If $|\psi\rangle$ and $|\phi\rangle$ are two pure states, swap test can be used to calculate $|\langle\psi|\phi\rangle|^2$, which is also a valid distance measure between the two states, known as quantum Fidelity (Jozsa [12]; Schumacher [13])."

"The target image is encoded in a quantum state, and passed as one of the inputs in the swap test circuit."

"We have used the 7-qubit real quantum processor ibm_lagos, which is compatible for encoding two 3-qubit states, and had minimum gate noise and readout error compared to all the 7-qubit real devices at the time of generating the data."

"Due to the cloud inaccessibility to higher dimensional systems, we could not check the swap test for four qubit states."

"In Garcia-Escartin and Chamorro-Posada [14], the authors showed that an equivalent circuit for swap test can be built without using the auxiliary qubit, at the cost of measuring all the qubits used to encode the states."

Destructive Swap Test for Pattern Recognition

"We present a comparative study between the swap test and destructive swap test in real quantum processors as well as in the ideal qasm_simulator provided by IBMQ."

"For the three-qubit images, this error is around 31%, which is still an improvement over the swap test."

"To perform destructive swap test for larger images, we need access to higher dimensional real quantum processors."

"Due to the impossibility of completely removing the presence of noise, the real IBMQ processors are subject to gate errors, state preparation and measurement errors, and readout errors."

"We vary p from 0.05 to 1.05 and calculate the overlap between two identical 2×2 image by destructive swap test."

"From the figure, the depolarizing noise in single qubit gates is more detrimental than the depolarizing noise in two-qubit gates for destructive swap test."

Pattern Recognition in Grayscale Images

"It is important to investigate whether the destructive swap test can efficiently identify patterns in grayscale/RGB images."

"Irrespective of the pixel values, we can encode these images in a quantum state using QPIE method, and then apply the destructive swap test to measure the overlap between them."

"Each image is 28×28 pixels, containing a handwritten digit between 0 and 9."

"As our target image, we pick a particular image containing '0' as the pattern, and compare it with a set of other images with different numbers."

"This is also due to the higher complexity of the images, as it is clear from the pictures that the pixels in these images encode more shades of gray compared to MNIST numbers."

Experimental Demonstration of Destructive Swap Test in Diamond NVs

"The IBMQ processors are based on superconducting qubit technology."

"To test the efficiency of destructive swap test in a different quantum system, we use two qubits in diamond NVs (Wrachtrup and Jelezko [15]; Suter and Jelezko [16]) to demonstrate the performance of destructive swap test."

"We choose the electron spin in states with $m_S = 0$ and $m_S = -1$ as qubit 1, and ^{13}C spin as qubit 2, where m_S denotes quantum number for the electron spin."

Supervised Learning with Destructive Swap Test

"To differentiate between these basis vectors and the rest, we apply CCNOT gates such that each pair of the target and reference state qubit acts as the control qubits, and the auxiliary qubit flips its state when both of the former are in state $|1\rangle$."

"The total state after applying the CCNOT gate can be written as: Thus, the qubit keeps the memory of the reference images, whereas the auxiliary qubits keep record of the basis vectors contributing to success and failure probabilities."

"To label the reference states, we use the basis states $\{|00\rangle,|01\rangle,|10\rangle,|11\rangle\}$ of a two-qubit system and prepare the following superposition: Since the target image is $|S_1\rangle$, we expect that at the end of the protocol described above, the joint measurement on the two-qubit system and the auxiliary qubit will give highest probability corresponding to the state $|110\rangle$."

Conclusion

"This naturally implies that swap test, which calculates the closeness between two quantum states, is a plausible tool to identify similar patterns in quantum images."

"Our results showed that for the typical values of the gate errors present in IBMQ quantum machines, the output of swap test becomes completely noisy corresponding to three or more qubits."

"With access to larger quantum processors, it will be interesting to see whether the destructive swap test still remains successful against the noise for higher dimensional states."

"To apply destructive swap test to larger images, we divided them into smaller segments which can be encoded using the 5 or 7 qubit quantum devices."

"If it is possible to encode the full quantum states using the real IBMQ processors with suitably controlled noise, the destructive swap test is without any doubt a potentially useful tool for pattern recognition, as clear from our results in this work."

Amplitude Transformed Quantum Convolutional Neural Network [17]

This is a machine-generated summary of:

Di, Shiqin; Xu, Jinchen; Shu, Guoqiang; Feng, Congcong; Ding, Xiaodong; Shan, Zheng: Amplitude transformed quantum convolutional neural network [17].

Published in: Applied Intelligence (2023).

Link to original: https://doi.org/10.1007/s10489-023-04581-w

If you want to cite the papers, please refer to the original.

For technical reasons we could not place the page where the original quote is coming from.

Abstract-Summary

"With the rapid development of quantum neural networks (QNN), several quantum simulations of convolutional neural networks (CNN) have been proposed."

"Google has proposed three quantum convolutional neural network (QCNN) models, but its purely QCNN model suffers from slow convergence and low training efficiency."

"We propose an Amplitude Transformed Quantum Convolutional Neural Network (ATQCNN)."

"Experiments show that our model achieves 100% and 97.92% accuracy and faster convergence on the quantum cluster state and CICMalDroid2020 datasets compared to the purely QCNN proposed by Google."

Introduction

"The entanglement, superposition, and unitary evolution of quantum computing not only can solve the computational efficiency problem of machine learning but also be used to develop more intelligent machine learning algorithms."

"In 2019, a Quantum Convolutional Neural Network circuit model was proposed by Cong and others [18]."

"There are still some problems in the research related to QCNN. For example, the purely quantum convolutional neural network model proposed by Google in the literature [19] has the longest training time, slow convergence rate, and low training efficiency compared with other models."

"Compared with the classical neural network, the quantum machine learning algorithm based on PQCs has more substantial expressive and computing ability due to the characteristics of quantum mechanics [20]."

"Our aim is to construct QCNN models that can efficiently represent the solution space with less quantum cost by parameterized quantum circuits (PQCs)."

Preliminaries

"We introduce basic concepts of quantum computation, parameterized quantum circuits (PQCs), quantum state encoding and the QCNN models proposed by Google and related experiments."

"Any unitary transformation can be decomposed into some set of basic quantum gates [], so the states of qubits can be manipulated by quantum gates."

"The single-qubit rotating gates contains parameters that can be iteratively optimized, and the parameters determine the effect exerted by the quantum gate on the quantum state and the measurement results of the quantum state."

"In the first purely quantum CNN, convolution and pooling layers are applied iteratively until the system size is reduced to 1 qubit."

"Unlike the purely quantum CNN, both hybrid models perform only one quantum convolution and pooling operation, after which the measured results are fed into the classical neural network and get the final output of the model as the classification result."

Amplitude Transformed Quantum Convolutional Neural Network

"Compared with the purely quantum CNN proposed by Google, We added the design of a quantum fully connected layer to the model."

"The two models have different compositional structures of the quantum convolution and quantum pooling layers."

"After extracting features layer by layer, the expected value is obtained by quantum measurement of a specific qubit and the loss function is calculated."

"The l denotes the l-th quantum convolutional and pooling layer and the i denotes the i-th unitary operation in each layer, and the N is the number of qubits in the model."

"The primary purpose of convolutional layers is to extract features from input data using feature maps, and quantum convolution has the advantage of enhancing maps [21]."

"Same as the quantum convolutional layer, it has the same parameters within a layer."

Experiments and Results

"The difference is that the compared with Google's purely quantum CNN, ATQCNN has an additional quantum FC layer."

"Google's purely quantum CNN contains 63 parameters to train, while ATQCNN only contains 46 parameters."

"The average training time of a single sample of the Google model is about 55 ms on an analog processor, the ATQCNN model can be improved by about 40%, reaching about 35 ms."

"Noted that the 8-qubit ATQCNN contains 46(14*3 + 4) parameters, but Google's two hybrid models have 69(21 + 6*8) and 175(21*3 + 8*14) parameters, respectively, both exceeding the ATQCNN."

"The ATQCNN model is used for a simple malicious code classification task to demonstrate the potential capabilities of quantum machine learning."

"ATQCNN achieves 97.92% accuracy, Google's two hybrid models have an accuracy of 98.85%, but Google's pure quantum model only achieves 87.5%."

Conclusion

"We design low-depth parameterized quantum circuits with only two quantum bits interacting, and construct a QCNN framework with lower depths, fewer parameters and global correlation."

"Compared to Google's purely quantum model, ATQCNN converges faster and more stably, and improves training efficiency by 35%. In particular, the required parameters and depth of ATQCNN are in reduced by about 27% for the same scale of qubits."

"It is more suitable for handling quantum data and classical data with low feature dimensionality, and performs poorly on datasets with high feature dimensionality."

"When we design and optimize quantum circuits, we consider fewer parameters and lower depths, ignoring the expressiveness of the model."

Genetic Algorithm for Searching Bipolar Single-Flux-Quantum Pulse Sequences for Qubit Control [22]

This is a machine-generated summary of:

Bastrakova, M. V.; Kulandin, D. S.; Laptyeva, T.; Vozhakov, V. A.; Liniov, A. V.: Genetic Algorithm for Searching Bipolar Single-Flux-Quantum Pulse Sequences for Qubit Control [22].

Published in: Lobachevskii Journal of Mathematics (2023).

Link to original: https://doi.org/10.1134/s1995080223010043

Copyright of the summarized publication:

If you want to cite the papers, please refer to the original.

For technical reasons we could not place the page where the original quote is coming from.

Abstract-Summary

"They require implementation of energy efficient qubit state control scheme."

"A promising approach is the use of superconducting digital circuits operating with single-flux-quantum (SFQ) pulses."

"The duration of SFQ pulse control sequence is typically larger than that of conventional microwave drive pulses but its length can be optimized for the system with known parameters."

"We introduce a genetic algorithm for unipolar or bipolar SFQ control sequence search that minimize qubit state leakage from the computational subspace."

Introduction

"One of them is the use of digital devices of superconducting electronics [7–11] using sequences of control pulses with a wide spectrum—Single Flux Quantum (SFQ) pulses."

"This method has successfully proven the possibility to implement the single-qubit operations with regular sequences [23] and optimized pulse sequences of SFQ pulses (SCALLOPS approach) [24]."

"It was possible to demonstrate the implementation of short (12–20 ns) single-qubit operations with the performance compared to the technique using more common room temperature microwave equipment."

"We have shown that it is possible to accelerate quantum operations on a transmon qubit using bipolar sequences of the SFQ pulses."

"We have proposed an original genetic algorithm for searching for the bipolar sequences aimed at minimizing leakage from the computational qubit subspace (ground and first excited states)."

"Section 2 outlines the model of transmon qubit and its control with SFQ pulses [22]."

"The genetic algorithms for Searching SFQ pulse sequences are described in Section 3 [22]."

Physical Model of the Transmon Qubit

"We are considering an ordinary transmon qubit [25]."

"This leads to a strong decrease in the charge fluctuations influence on the qubit state."

Bipolar Sequence Search Algorithms

"Bipolar SCALLOP sequences are optimized with the genetic-like algorithm aimed at the minimization of leakage."

"The second one searches for the optimal subsequence of a given length, the repetition of which a certain number of times (we consider from 1 to 35 repetitions) gives the minimum leakage (see "Score Current Population" of the algorithm)."

"2 of them with the best fitness function values become parents."

"Add Children to Next Population: Adjust New Population Two individuals of the new generation with the worst fitness function values are replaced by two individuals of the previous generation with the best."

"Save Results The best sequence, the values of the obtained rotation angle and fidelity, and the running time of the algorithm are saved."

"When run with the input given in article [24], it found sequences with comparable angle rotation and fidelity values."

SFQ Pulse Sequences Search Results

"The use of subsequences does not always provide an acceptable rotation accuracy."

"We employed computing nodes with the following configuration: 2x Intel Xeon Silver 4310 T CPU (20 cores, 2.30 GHz), 64 GB RAM, OS CentOS 7.9."

"The low efficiency of such parallelization is due to the very small amount of data used and frequent switching between parallel and serial parts of code."

"Several computations for different sets of parameters, but the same lengths of sequences/subsequences, can be run on the same node in serial mode with binding each computation to a separate core."

"We obtained a parallelization efficiency of about 80% for the sequence search and 70% for the subsequence search while running 20 calculations simultaneously on 20 physical cores."

Conclusions and Future Work

"The simulation of coherent control of the transmon qubits using irradiation with sequences of SFQ pulses of different polarities."

"We selected a set of input parameters (similar to those presented in [24]) consisting of 21 qubit frequencies controlled by a single global clock at 25 GHz generator frequency."

"We have proposed the genetic algorithm for generating bipolar sequences."

"Such a control method can be used to implement two-qubit gates, similar to the microwave cross-resonance technique [26]."

Entanglement Concentration of Multi-Qubit Entangled States: An IBM Quantum Experience [27]

This is a machine-generated summary of:

Ram, Jagat; Dutt, Dev; Dhiman, S. K.; Behera, Bikash K.; Panigrahi, Prasanta K.: Entanglement concentration of multi-qubit entangled states: an IBM quantum experience [27].

Published in: Quantum Studies: Mathematics and Foundations (2023).

Link to original: https://doi.org/10.1007/s40509-023-00298-0

If you want to cite the papers, please refer to the original.

For technical reasons we could not place the page where the original quote is coming from.

Abstract-Summary

"We have applied the entanglement concentration protocols to partially entangled subsets of classes of highly entangled multi-qubit states, created on IBM quantum computer."

"The success of the protocol is analysed through calculating the success probability obtained from the IBM quantum computer and simulator with which the protocol can generate maximally entangled states."

Introduction

"As the interaction of quantum states with environment cannot be avoided, yet they can be processed to retrieve back as the original maximally entangled ones."

"In entanglement concentration process, the pure imperfect or partially entangled states get turned into the maximally entangled ones like Bell pairs with the help of local operations and classical communications (LOCC)."

"Decoherence is very difficult to tackle because the interaction of quantum systems with environment is unavoidable and consequently, the maximally entangled states get changed into the noisy ones, which are unsuitable for quantum information processing tasks."

"We are using these two ECP protocols for certain subsets of multi-qubit maximally entangled states of the following types: two-qubit Z-like states [28] and three-qubit Cluster-like state [29–31], three qubit cat-like state and a five-qubit AME-like (Absolutely Maximally Entangled) state [32]."

"Our main objective is to test these two protocols on IBM quantum computer and to observe its efficiency in generating the maximally entangled state using these two ECPs."

Theory

"On applying suitable unitary operations on the first qubit, we can get the desired maximally entangled state."

"Z-states are the n-qubit highly entangled multipartite states proposed in Ref. [28]."

"Applying the proposed ECP to the above state with the help of following non maximally entangled Bell-like state aswe get the final state asNow, performing a Bell basis measurement (BBM) on the first two-qubits of the above state and applying the suitable unitary operations, we obtain the desired maximally entangled two-qubit Z state."

"For our purpose, we consider the non-maximally entangled three-qubit cluster state."

Implementations on IBM Platforms

"A deviation of more than 5 appearing in the ECP can be justified as due to the error rates arising with the increasing number of qubits and gates in the quantum circuit."

"The faithful generation of quantum states means that a quantum circuit which comprises of arbitrary number of qubits for the accomplishment of certain computational or information processing tasks, the IBM quantum processor or any other NISQ [33] device should generate all the basis states efficiently with the desired probabilities."

"The incompetency of an NISQ devise to generate multi-qubit maximally entangled states affects the efficient functioning of a quantum protocol executed on the device."

"In this quantum protocol, the multi-qubit entangled states are mapped to an ansatz state (a many-body entangled wave function) which is the key ingredient of VQE algorithm [34]."

Conclusion

"The ability implies the success probability of the entanglement concentration protocol (ECP) with which it can filter out MES from the ensembles of partially entangled or imperfect quantum states."

"It can be asserted that the ECP assisted quantum circuit for any quantum protocol or any kind of quantum algorithm which is comprised of partially entangled or noisy states with a lesser number of qubits can be simulated on the IBMQ processor with a large fidelity or greater efficacy."

"For a quantum system involving more than three qubits like the five-qubit AME state, the inevitable errors arising in the quantum circuits for the state preparation alone (This means that when the pure maximally entangled five-qubit AME state is prepared on IBMQ processor), are so large that even the use ECPs cannot filter out the maximally entangled states from the IBM quantum processor."

Low-Loss Interconnects for Modular Superconducting Quantum Processors [35]

This is a machine-generated summary of:

Niu, Jingjing; Zhang, Libo; Liu, Yang; Qiu, Jiawei; Huang, Wenhui; Huang, Jiaxiang; Jia, Hao; Liu, Jiawei; Tao, Ziyu; Wei, Weiwei; Zhou, Yuxuan; Zou, Wanjing; Chen, Yuanzhen; Deng, Xiaowei; Deng, Xiuhao; Hu, Changkang; Hu, Ling; Li, Jian; Tan, Dian; Xu, Yuan; Yan, Fei; Yan, Tongxing; Liu, Song; Zhong, Youpeng; Cleland, Andrew N.; Yu, Dapeng: Low-loss interconnects for modular superconducting quantum processors [35].

Published in: Nature Electronics (2023).

Link to original: https://doi.org/10.1038/s41928-023-00925-z

If you want to cite the papers, please refer to the original.

For technical reasons we could not place the page where the original quote is coming from.

Abstract-Summary

"We use these interconnects to link five quantum modules with intermodule quantum state transfer and Bell state fidelities of up to 99%."

"To benchmark the overall performance of the processor, we create maximally entangled, multiqubit Greenberger–Horne–Zeilinger states."

"The generated intermodule four-qubit Greenberger–Horne–Zeilinger state exhibits 92.0% fidelity."

"We also entangle up to 12 qubits in a Greenberger–Horne–Zeilinger state with 55.8 ± 1.8% fidelity, which is above the genuine multipartite entanglement threshold of 1/2."

Main

"The near-term development of processors with thousands of qubits [36] for use in intermediate-scale applications [33] comes with a range of engineering challenges—including available wafer size, device yield and crosstalk [37, 38]—that limit the development of monolithic quantum processors."

"Modular quantum computing schemes may be the most viable approach to scaling up to very large qubit numbers in the near term [38]."

"With the development of quantum information science, the need for low-loss cables/waveguides has become imperative [38]."

"Several recent experiments have demonstrated the connection of two superconducting quantum modules using niobium–titanium (NbTi) superconducting coaxial cables [37, 39–44]."

"In this Article, we report coherent interconnects using low-loss coaxial cables made from pure Al, with Al wire-bonded connections to qubit processors."

"Although $T_c = 1.2$ K of Al is much lower than that of NbTi (9.7 K), the Q_{cb} value of our cables at ~10 mK reaches 4.2×10^6, which is equivalent to a loss of 0.15 dB km^{-1}."

Modular Quantum System Architecture

"To suppress dephasing noise and retain some frequency tunability, we use asymmetric Josephson junctions with $\alpha = E_{J1}/E_{J2} = 5.3$, where E_{J1} and E_{J2} are the Josephson energies of the two qubit junctions, giving a qubit frequency-tuning range of ~4.2 to ~5.1 GHz."

"The qubit frequency-tuning range here allows access to the $m = 10$ mode at 4.450 GHz and the $m = 11$ mode at 4.885 GHz."

"This scheme balances the loss in the communication mode as well as in the qubits [45]."

"We note that although stronger coupling can be achieved here, extrinsic qubit loss could emerge as the coupling increases, exposing the qubits to loss channels at the wire-bond joint [44]."

Benchmarking the Modular Processor Performance

"Greenberger–Horne–Zeilinger (GHZ) states [46] are a fundamental resource in fault-tolerant quantum computation and quantum communication; as they are highly susceptible to errors and decoherence, they provide a sensitive benchmark for our processor."

"As quantum state tomography requires measurements and computational resources that grow exponentially with the number of qubits, a full characterization of these two states is impractical [47]."

"The density matrix ρ of an ideal N-qubit GHZ state has only four non-zero elements: The two diagonal elements $\rho_{0\ldots0,0\ldots0}$ and $\rho_{1\ldots1,1\ldots1}$, and two conjugate off-diagonal elements $\rho_{0\ldots0,1\ldots1}$ and $\rho_{1\ldots1,0\ldots0}$, allowing one to evaluate the state fidelity in an easier way."

"The oscillation period of 2π/N is a unique feature of the N-qubit GHZ state."

Conclusions

"We have demonstrated their use in linking together quantum modules to create a large-scale quantum computing system."

"We have shown high-fidelity intermodule QST, entanglement generation and large-scale entangled states linking multiple quantum modules, rivalling those of monolithic designs."

"Our approach could be used to scale up modular superconducting quantum processors to large sizes and explore sophisticated quantum computation and simulation tasks."

Grammar-Aware Sentence Classification on Quantum Computers [48]

This is a machine-generated summary of:

Meichanetzidis, Konstantinos; Toumi, Alexis; de Felice, Giovanni; Coecke, Bob: Grammar-aware sentence classification on quantum computers [48].

Published in: Quantum Machine Intelligence (2023).

Link to original: https://doi.org/10.1007/s42484-023-00097-1

If you want to cite the papers, please refer to the original.

For technical reasons we could not place the page where the original quote is coming from.

Abstract-Summary

"In the area of quantum computing (QC), with the steady growth of quantum hardware and notable improvements towards implementations of quantum algorithms, we are approaching an era when quantum computers perform tasks that cannot be done on classical computers with a reasonable amount of resources."

"We work with the Categorical Distributional Compositional (DisCoCat) model of natural language meaning, whose underlying mathematical underpinnings make it amenable to quantum instantiations."

"Work on fault-tolerant quantum algorithms has already demonstrated potential quantum advantage for NLP, notably employing DisCoCat."

"We focus on the capabilities of noisy intermediate-scale quantum (NISQ) hardware and perform the first implementation of an NLP task on a NISQ processor, using the DisCoCat framework."

"Our novel QNLP model shows concrete promise for scalability as the quality of the quantum hardware improves in the near future and solidifies a novel branch of experimental research at the intersection of QC and AI."

Introduction

"A prominent approach attempting this merge is the Distributional Compositional Categorical model of natural language meaning (DisCoCat) (Coecke and others [49]; Grefenstette and Sadrzadeh [50]; Kartsaklis and Sadrzadeh [51]), which pioneered the paradigm of combining explicit grammatical (or syntactic) structure with distributional (or statistical) methods for encoding and computing meaning (or semantics)."

"Collection of ingredients, there organically emerges the interdisciplinary field of quantum natural language processing (QNLP), a research area still in its infancy (Zeng and Coecke [52]; O'Riordan and others [53]; Wiebe and others [54]; Bausch and others [55]; Chen [56]), combines NLP and QC and seeks novel quantum language model designs and quantum algorithms for NLP tasks."

"The categorical formulation of quantum theory is known as Categorical Quantum Mechanics (CQM) (Abramsky and Coecke [57]) and it becomes apparent that the string diagrams describing CQM are tensor networks endowed with a graphical language in the form of a rewrite-system (a.k.a.)"

"The language of string diagrams places syntactic structures and quantum processes on equal footing, and thus allows the canonical instantiation of grammar-aware quantum models for NLP."

The Model

"In string diagrams representing pregroup-grammar reductions, words are represented as states and pairwise type-reductions are represented by a pattern of nested cup-effects (wires bent in a U-shape), and identities (straight wires)."

"Given a string diagram resulting from the grammatical reduction of a sentence, we can instantiate a model for natural language processing by giving semantics to the string diagram."

"As we have described in Meichanetzidis and others [58], the string diagram of the syntactic structure of a sentence σ can be canonically mapped to a PQC $C_\sigma(\theta_\sigma)$ over the parameter set θ. The key idea here is that such circuits inherit their architecture, in terms of a particular connectivity of entangling gates, from the grammatical reduction of the sentence."

"In 'classical' implementations of DisCoCat, where the semantics chosen in order to realise a model is in terms of tensor networks, a sentence diagram represents a vector which results from a tensor contraction."

Classification Task

"The dataset or 'labelled corpus' $K = \{(D_\sigma, l_\sigma)\}_\sigma$ is a finite set of sentence-diagrams $\{D_\sigma\}_\sigma$ constructed from a finite vocabulary of words V. Each sentence has a binary label $l_\sigma \in \{0,1\}$."

"We then quantify the performance by the training and test errors e_Δ and e_E, as the proportion of labels predicted incorrectly: This supervised learning task of binary classification for sentences is a special case of question answering (QA) (de Felice and others [59]; Chen and others [60]; Zhao and others [61]); questions are posed as statements and the truth labels are the binary answers."

"After training on Δ, the model predicts the answer to a previously unseen question from E, which comprises sentences containing words all of which have appeared in Δ. The optimisation is performed over the parameters of all the sentences in the training set $\theta = \cup_{\sigma \in \Delta} \theta_\sigma$."

"Recall that a given sentence-circuit does not necessarily involve the parameters of every word."

Discussion and Outlook

"Our DisCoCat-based QNLP framework is naturally generalisable to accommodate mapping sentences to quantum circuits involving mixed states and quantum channels."

"As also stated above, it is possible to define functors in terms of hybrid models where both neural networks and PQCs are involved, where heuristically one aims to quantify the possible advantage of such models compared to strictly classical ones."

"Looking beyond the DisCoCat model, it is well-motivated to adopt the recently introduced DisCoCirc model (Coecke [62]) of meaning and its mapping to PQCs (Coecke and others [63]), which allows for QNLP experiments on text-scale real-world data in a fully-compositional framework."

"This interaction structure, viewed as a process network, can again be used to instantiate models in terms of neural networks, tensor networks, or quantum circuits."

"Entities are modelled as density matrices carried by wires and their modifiers as quantum channels."

Experimental Quantum Adversarial Learning with Programmable Superconducting Qubits [64]

This is a machine-generated summary of:

Ren, Wenhui; Li, Weikang; Xu, Shibo; Wang, Ke; Jiang, Wenjie; Jin, Feitong; Zhu, Xuhao; Chen, Jiachen; Song, Zixuan; Zhang, Pengfei; Dong, Hang; Zhang, Xu; Deng, Jinfeng; Gao, Yu; Zhang, Chuanyu; Wu, Yaozu; Zhang, Bing; Guo, Qiujiang; Li, Hekang; Wang, Zhen; Biamonte, Jacob; Song, Chao; Deng, Dong-Ling; Wang, H.: Experimental quantum adversarial learning with programmable superconducting qubits [64].

Published in: Nature Computational Science (2022).

Link to original: https://doi.org/10.1038/s43588-022-00351-9

If you want to cite the papers, please refer to the original.

For technical reasons we could not place the page where the original quote is coming from.

Abstract-Summary

"Recent theoretical works show that, similar to traditional classifiers based on deep classical neural networks, quantum classifiers would suffer from adversarial perturbations as well."

"We report an experimental demonstration of quantum adversarial learning with programmable superconducting qubits."

"We demonstrate that these well-trained classifiers (with testing accuracy up to 99%) can be practically deceived by small adversarial perturbations, whereas an adversarial training process would substantially enhance their robustness to such perturbations."

Main

"It has been shown theoretically that quantum classifiers are likewise highly vulnerable to adversarial examples, independent of the learning algorithms and regardless of whether the input data are classical or quantum [65]."

"We generate adversarial examples through a classical optimizing procedure and unambiguously show that they can deceive the trained quantum classifiers with a high confidence level."

"To mitigate such vulnerability, we further demonstrate that, through adversarial training, the quantum classifiers will be immune to adversarial perturbations generated by the same attacking strategy."

"This demonstrates the vulnerability of quantum classifiers to adversarial perturbations in categorizing quantum states."

"We optimized the device fabrication and controlling process to minimize hardware noises, which is crucial for both implementation of the large-scale quantum classifiers and demonstration of their vulnerability to adversarial perturbations."

"The experimental demonstration of quantum adversarial examples for unsupervised learning and other types of quantum classifier [66] seems more technically sophisticated and remains unattainable."

Methods

"We consider classification tasks in the setting of supervised learning [67–69], where we train quantum classifiers with pre-labeled data samples by minimizing the following loss function iteratively:Here, x denotes a training sample, h(x; θ) represents the hypothesis function determined by the quantum classifier with variational parameters denoted collectively as θ, a is the one-hot encoding of the labels, and g_k denotes the probability for the kth category obtained from measuring the quantum classifier."

"Because the parameters in our case are all encoded in the angles of single-qubit Pauli-rotation gates, we can optimize the quantum classifier with gradients obtained from measurements."

"After the training process, the quantum classifier will typically be able to assign labels to data samples outside the training set with high accuracy."

"In the training (test) set with size 500 (100), we numerically generate a corresponding adversarial example on a classical computer aiming to mislead the well-trained classifier to make an incorrect prediction."

Neural Error Mitigation of Near-Term Quantum Simulations [70]

This is a machine-generated summary of:

Bennewitz, Elizabeth R.; Hopfmueller, Florian; Kulchytskyy, Bohdan; Carrasquilla, Juan; Ronagh, Pooya: Neural Error Mitigation of Near-Term Quantum Simulations [70].

Published in: Nature Machine Intelligence (2022).

Link to original: https://doi.org/10.1038/s42256-022-00509-0

If you want to cite the papers, please refer to the original.

For technical reasons we could not place the page where the original quote is coming from.

Abstract-Summary

"We introduce neural error mitigation, which uses neural networks to improve estimates of ground states and ground-state observables obtained using near-term quantum simulations."

"To demonstrate our method's broad applicability, we employ neural error mitigation to find the ground states of the H_2 and LiH molecular Hamiltonians, as well as the lattice Schwinger model, prepared via the variational quantum eigensolver."

"Neural error mitigation is agnostic with respect to the quantum state preparation algorithm used, the quantum hardware it is implemented on and the particular noise channel affecting the experiment, contributing to its versatility as a tool for quantum simulation."

Main

"Examples include the implicit characterization of noise models and how they affect estimates of the desired observables, specific knowledge of the state subspaces in which the prepared quantum state ought to reside and the characterization and mitigation of the sources of noise on individual components of the quantum computation, such as single- and two-qubit gate errors, as well as state preparation and measurement errors."

"We introduce a QEM strategy named neural error mitigation (NEM), which uses neural networks to mitigate errors in the approximate preparation of the quantum ground state of a Hamiltonian."

"We performed neural quantum state (NQS) tomography (NQST) to train an NQS ansatz to represent the approximate ground state prepared by a noisy quantum device using experimentally accessible measurements."

"The overhead of error mitigation in NEM is shifted from quantum resources (that is, performing additional quantum experiments and measurements) to classical computing resources for machine learning."

Results

"We highlight the performance of NEM on the experimental preparation of the ground states of LiH at different bond lengths using IBM's five-qubit chip, IBMQ-Rome."

"The median performance of NEM improves the ground-state estimation of H_2 and LiH to chemical accuracy and low infidelities for all bond lengths."

"We demonstrated the performance of NEM by applying it to the approximate ground state of the lattice Schwinger model obtained by numerically simulating a VQE algorithm for $N = 8$ sites, with single-qubit depolarizing noise with probability $\lambda = 0.001$ applied after each rotation and entangling operation."

"The ability to obtain precise estimations of these physical properties can be explained by the accurate representation of the ground-state wavefunction captured by the NEM NQS."

"The VQE algorithm was simulated on a classical computer for system sizes up to $N = 16$, and NEM was applied to the resulting states."

Discussion

"The error mitigation strategy developed here demonstrates substantial improvements in estimations of ground states and ground-state observables obtained from two example classes of near-term quantum simulations, independent of the quantum device and noise channels."

"Given its low quantum overhead, NEM could be a powerful asset for error mitigation of near-term quantum simulations."

"The accurate final representation of the ground-state wavefunction is the reason why NEM is able to accurately reconstruct and improve estimations of complex observables such as energy, order parameters and entanglement entropy without requiring additional quantum resources."

"By combining VQE, which uses a parametric quantum circuit as an ansatz, and NQST and VMC, which use neural networks as an ansatz, NEM brings together two families of parametric quantum states and three optimization problems over their loss landscapes [71–73]."

"Our work raises the question of the nature of the relationships between these families of states, their loss landscapes and quantum advantage."

Methods

"We represent the quantum state $|\Psi\rangle$ with a Transformer neural network that takes as input a bitstring $s = (s_1, \ldots, s_N) \in \{0, 1\}^N$, describing a computational basis state $|s\rangle$, where N is the number of qubits."

"The method usually proceeds by gradient-based optimization of the energy, where the energy and its partial derivatives with respect to the ansatz parameters are estimated using Monte Carlo samples drawn from the classical variational wavefunction."

"The variational circuit was then optimized using Qiskit's implementation of simultaneous perturbation stochastic approximation (SPSA) [74] for 250 iterations to obtain an estimation for the ground-state energy of H_2 and LiH. Each SPSA iteration required two energy evaluations."

"Evolution with this Hamiltonian preserved the symmetries of the lattice Schwinger model to first-order terms in J/B. Only half of the parameters in each single-qubit rotation layer were independent, as required by the symmetries, giving $\varphi_j = -\varphi_{N+1-j}$ for $j \in \{N/2 + 1, \ldots, N\}$."

Mixer-Phaser Ansätze for Quantum Optimization with Hard Constraints [75]

This is a machine-generated summary of:

LaRose, Ryan; Rieffel, Eleanor; Venturelli, Davide: Mixer-phaser Ansätze for quantum optimization with hard constraints [75].

Published in: Quantum Machine Intelligence (2022).

Link to original: https://doi.org/10.1007/s42484-022-00069-x

If you want to cite the papers, please refer to the original.

For technical reasons we could not place the page where the original quote is coming from.

Abstract-Summary

"The ansätze are inspired by mixing and phase separation in the QAOA, and also motivated by compilation considerations with the aim of running on near-term superconducting quantum processors."

"For the parameter setting strategies and evaluation metric used, the average performance achieved by the QAOA is effectively matched by the one obtained by a" mixer-phaser" ansatz that can be compiled in less than half-depth of standard QAOA on most superconducting qubit processors."

Introduction

"The algorithms following the ansatz alternate p times between unitary operators chosen from a one-parameter family of phase separation operators and operators chosen from a one-parameter family of mixing operators."

"In Hadfield and others [76], the QAOA approach was generalized to the Quantum Alternating Operator Ansatz, considering unitary layers that are not necessarily linked to local Hamiltonian evolution."

"Multi-qubit mixing operators were introduced in lieu of the X-rotations when applying the QAOA to hard-constrained optimization in order to restrict the search to the feasible subspace, the space of valid configurations obeying the hard constraints."

"We introduce a new ansatz that combines the mixing and phase-separation operators into a more general two-parameter family of operators."

"Our numerical simulations show that for the problems studied, in the noiseless case, QAMPA performs almost as well as standard QAOA in parameter regimes that are achievable in current hardware, and thus is expected to have advantages under noise given its reduced depth."

Background and Prior Work

"Many techniques have been developed to optimize gate synthesis and qubit routing (i.e. compilation) for algorithms to be run on noisy intermediate scale quantum (NISQ) devices featuring a sparse native gate set."

"Although experimental QAOA work with XY mixers has still not appeared in the literature, numerical analyses predict long circuit durations for problems with XY mixers that are not encouraging for very near-term hardware (Do and others [77])."

"In that approach, QAOA was used as a form of a quantum neural network that needs to be trained to act as an optimization solver, but there is concern as to how well this method, with its many parameters, would work in general."

"Most of the works on QAOA feature exclusively single qubit gates as mixers."

"Only a few works have discussed the performance of the QAOA using XY -mixers."

QAMPA: Quantum Alternate "Mixer-Phaser" Ansatz

"Following the literature, we construct the QAOA and QAMPA gates using XY mixers (Hadfield and others [76]; Wang and others [78]) as: The initial state could be taken to be an equal superposition of all solutions with κ variables set to 1 on the qubit registers."

"Choosing a mixer with sparser connectivity between various terms might lead to shorter circuits, but the compilation advantage of using QAMPA versus QAOA is maximal if we use the same graph for both phase-separation and mixing operations."

"(Hadfield and others [76]; Egger and others [79]), the initialization procedure could be possibly substituted by a simpler to realize superposition of feasible states or a classical warm start candidate followed by a mixing round in QAOA."

"The routing requirements to schedule gates between all possible pairs of qubits depend on the underlying topology where swap operations can be performed."

Numerical Evaluation

"To address this goal, our target performance metric will be the expected value of the objective function when the best result in R runs is selected (Kim and others [80]): where $|k\rangle$ is a feasible state whose normalized objective function value is: with ϵ_0, $\epsilon^\star$ being respectively the minimum and the maximum of the objective function spectrum of values."

"The protocol aims to identify good angles for both QAOA and QAMPA at level p + 1 using the information of the best found at level p. It starts with a random generation of W_0 pairs of angles that are then used as an initialization for each run of the optimizer."

"What is observed is that the standard QAOA approach (which performs only slightly better than QAMPA, as we recall) gives the best performance compared against all other tested variations, and it is always beneficial to include a proportional factor multiplying γ_p, for each gate between qubits n and m, corresponding to the objective function coefficients J_{nm} in the circuit ansatz."

Discussion and Conclusions

"While theoretical frameworks to estimate the impact of noise in circuits featuring XY gates are being developed (Streif and others [81]), ultimately only the experimental tests on quantum hardware will be able to provide a full picture of performance, including identifying at what layer p the solver is most effective."

"Another recent experiment (Hashim and others [82]) shows that permuting the ordering of the qubits in the SWAP network and averaging over the results is reducing the systematic coherent errors, a technique that could be generalized to our case where the permutations are not equivalents, possibly helping the parameter setting and/or optimization performance."

"A rich ecosystem of ansätze that take into account hardware architectures, gate sets, and noise considerations for different types of NISQ processors will enable more rapid understanding of quantum optimization approaches for the NISQ era and beyond."

QDNN: Deep Neural Networks with Quantum Layers [83]

This is a machine-generated summary of:

Zhao, Chen; Gao, Xiao-Shan: QDNN: deep neural networks with quantum layers [83].

Published in: Quantum Machine Intelligence (2021).

Link to original: https://doi.org/10.1007/s42484-021-00046-w

If you want to cite the papers, please refer to the original.

For technical reasons we could not place the page where the original quote is coming from.

Abstract-Summary

"A quantum extension of classical deep neural network (DNN) is introduced, which is called QDNN and consists of quantum structured layers."

"It is proved that the QDNN can uniformly approximate any continuous function and has more representation power than the classical DNN."

"The QDNN still keeps the advantages of the classical DNN such as the non-linear activation, the multi-layer structure, and the efficient backpropagation training algorithm."

Introduction

"Approaches were proposed to build classical DNNs on quantum computers (Killoran and others [84]; Zhao and others [85]; Kerenidis and others [86])."

"The structure of classical DNNs is still used, and only some local operations are speeded up by quantum algorithms."

"Several quantum analogs of DNNs were proposed."

"Several NISQ quantum machine learning models based on PQCs, such as quantum generative adversarial networks, quantum circuit Born machine, and quantum kernel methods, were proposed (Lloyd and Weedbrook [87]; Dallaire-Demers and Killoran [88]; Liu and Wang [89]; Schuld and Killoran [90]; Havlíček and others [67]; Benedetti and others [91])."

"We introduce the quantum deep neural network (QDNN) which is a composition of multiple quantum neural network layers (QNNLs)."

"Unlike other approaches of quantum analogs of DNNs, our QDNN still keeps the advantages of the classical DNN such as the non-linear activation, the multi-layer structure, and the efficient backpropagation training algorithm."

"QDNN provides a new class of neural networks which can be used in near-term quantum computers and is more powerful than classical DNNs."

The QDNN

"The power of DNNs comes from the non-linearity of the activation function."

"All quantum gates are unitary matrices and hence linear."

"We build QNNLs using the hybrid quantum-classical algorithm scheme (McClean and others [92]), which is widely used in many NISQ quantum algorithms (Liu and Wang [89]; Liu and others [93])."

"After encoding the input data, we apply a linear transformation as the analog of the linear transformation in the classical DNNs."

"This part is natural on quantum computers because all quantum gates are linear."

"Notice that the QNNLs activate before affine transformations while classical DNNLs activate after affine transformations."

"The QNNLs can be naturally embedded in classical DNNs."

"In classical DNNs, the gradient of parameters in each layer is computed by the backpropagation algorithm (BP)."

"We can use Algorithm 2 to estimate the gradient in each quantum layer."

"Gradients can be backpropagated through the quantum layer, and QDNNs can be trained with the BP algorithm."

Representation Power of QDNNs

"We will show that QDNN can approximate any continuous function similar to the classical DNN."

"The universal approximation theorem ensures that DNNs can approximate any continuous function (Cybenko [94]; Hornik [95]; Leshno and others [96]; Pinkus [97])."

"Since the class of QDNNs is an extension of the class of classical DNNs, the universal approximation theorem can be applied to the QDNN trivially."

"The QDNN with only QNNLs can uniformly approximate any continuous function Now, let us consider the second case."

"The QDNN with QNNLs and affine layers can approximate any continuous function."

"Similar to the classical case, QDNNs with one quantum layer can approximate any continuous function (Hornik [95])."

"Under the hypothesis that quantum computers cannot be simulated efficiently by classical computers, which is generally believed, there exists a function represented by a QDNN which cannot be computed by classical circuits of polynomial size."

Experimental Results

"We will use QDNNs to conduct a numerical experiment for an image classification task."

"The goal of this QDNN is to recognize the digit in the image is either 0 or 1 as a classifier."

"We use three QNNLs in our QDNN, which will be called the input layer, the hidden layer, and the output layer, respectively."

"We use trainable QDNNs by adopting local Hamiltonians and small depth structure."

"Because of the small depth of the ansatz and the local Hamiltonian, the QDNN is trainable (Cerezo and others [98])."

"Because there are 30 parameters in the encoder, we set the last column of R_Z gates to be $R_Z(0)$."

"Because there are 20 parameters in the encoder, we set the last column of R_Z and R_X gates to be $R_Z(0)$ and $R_X(0)$."

"We use Adam optimizer (Kingma and Ba [99]) to update parameters."

Discussion

"We proved that QDNNs have more representation power than classical DNNs."

"The QDNN still keeps most of the advantages of the classical DNNs."

"Because the model is based on the hybrid quantum-classical scheme, it has the potential to be realized on NISQ processors."

"Since we use a classical simulator on a desktop PC for quantum computation, only QDNNs with a small number of qubits can be used and only simple examples can be demonstrated."

"With quantum computing resources, we can access exponential dimensional feature Hilbert spaces (Schuld and Killoran [90]) with QDNNs and only use polynomial-size parameters."

Efficient Parallelization of Tensor Network Contraction for Simulating Quantum Computation [100]

This is a machine-generated summary of:

Huang, Cupjin; Zhang, Fang; Newman, Michael; Ni, Xiaotong; Ding, Dawei; Cai, Junjie; Gao, Xun; Wang, Tenghui; Wu, Feng; Zhang, Gengyan; Ku, Hsiang-Sheng; Tian, Zhengxiong; Wu, Junyin; Xu, Haihong; Yu, Huanjun; Yuan, Bo; Szegedy, Mario; Shi, Yaoyun; Zhao, Hui-Hai; Deng, Chunqing; Chen, Jianxin: Efficient parallelization of tensor network contraction for simulating quantum computation [100].

Published in: Nature Computational Science (2021).

Link to original: https://doi.org/10.1038/s43588-021-00119-7

If you want to cite the papers, please refer to the original.

For technical reasons we could not place the page where the original quote is coming from.

Abstract-Summary

"We develop an algorithmic framework for contracting tensor networks and demonstrate its power by classically simulating quantum computation of sizes previously deemed out of reach."

"We then demonstrate applications of the simulation framework for aiding the development of quantum algorithms and quantum error correction."

"As tensor networks are widely used in computational science, our simulation framework may find further applications."

Main

"Claims of quantum superiority are based on the assumption that classical simulation cannot achieve a task that is easily achievable using quantum devices."

"We propose a highly optimized framework for classically simulating intermediate- to large-scale quantum computations, represented as tensor networks."

"Tensor network contraction has been one of the prominent choices for simulating quantum computation due to its high flexibility and expressive power; however, exact contraction of general tensor networks is a computationally hard problem with respect to the problem size: there are tensor networks for which an exact simulation would take exponential amount of time under well-established computational complexity assumptions [101, 102]."

"To demonstrate the usefulness and broad capabilities of the tensor network-based simulation framework, we apply it to both the studies of near-term quantum algorithms and fault-tolerant quantum computing."

Results

"To remedy these two problems, we propose a parallelization framework called index slicing that aims to divide a tensor network contraction task into many subtasks with identical tensor network structures such that the subtasks can be executed in parallel, each with a space complexity small enough to fit into a single computational unit."

"Throughout the results we report the complexity of contracting different tensor networks by a pair of numbers: The first one, called the computation cost, is the base ten logarithm of the number of total floating point operations (FLOPS), serving as a measure for the time complexity."

"We measure the performance of various simulation frameworks based on tensor network contraction with the contraction cost and an extrapolated running time that is based on actually running some of the many structurally identical subtasks created by index slicing."

Discussion

"For a tensor network with m indices associated with a hypergraph with tree width t and contraction width c, the sequential pairwise contraction achieves a time complexity of $O^*(2^t)$, whereas the space complexity is lower bounded by $\Omega^*(2^c)$ (although not necessarily simultaneously achievable)."

"It also remains open whether there exists a tensor network contraction algorithm that achieves both the relatively low time complexity of $O^*(2^t)$ and the space complexity of O(m), and it does not seem likely that a slicing-incorporated sequential pairwise contraction could achieve this limit."

"A better contraction order—together with index slicing—might be found through algorithmic refinements; however, it is known [101] that exact contraction of general tensor networks is a #P-hard problem, a computational complexity category for which no known efficient (subexponential time) algorithms exist."

"It has been noted that the good contraction schemes found by our algorithm, usually with relatively low time complexity, might not necessarily perform well on modern computational architectures."

Methods

"Finding the optimal contraction scheme (that is, a subset of indices to slice over and a sequential pairwise contraction order for the subtasks; identical in structure) is NP-hard; however, for large instances of tensor networks presented in this paper, a preprocessing heuristic finding near-optimal contraction schemes is often worthwhile as it makes a big difference in time/space complexities for the actual contraction task that considerably dwarfs the relatively short extra time spent on such preprocessing."

"We apply the following heuristics: The first one is a general local optimization method: take a connected subgraph of a contraction tree, which represents a series of contraction steps, with multiple intermediate outcomes as the input and a single output."

"This increases the overall unsliced cost (assuming that the original contraction tree is locally optimal), but at the same time reduces the slicing overhead via increasing the utility of the particular index."

Compressed and Canalized Emission of Quantum Emitters in MIM Nano-Cavities [103]

This is a machine-generated summary of:

Palermo, Giovanna; Lio, Giuseppe E.; Strangi, Giuseppe: Compressed and canalized emission of quantum emitters in MIM nano-cavities [103].

Published in: Quantum Studies: Mathematics and Foundations (2020).

Link to original: https://doi.org/10.1007/s40509-020-00231-9

If you want to cite the papers, please refer to the original.

For technical reasons we could not place the page where the original quote is coming from.

Abstract-Summary

"We numerically demonstrate that quantum emitters can interact coherently with a sub-wavelength plasmonic nano-cavity."

"This effect is observed in a cascade of metal–insulator–metal nanostructures (MIM), interlocked by a nano-cavity containing the quantum emitters, where the selective reshaping of the radiated field occurs."

"Our study pave the way for the implementation of cavity-mediated quantum gates and for the realization of scalable and tunable quantum systems."

Introduction

"The interaction between quantum emitters and photons is key for the development of quantum optical networks [104–106], superconducting quantum processors [107, 108], enabling quantum logic operations and entanglement generation by means of stationary quantum bits (qubits) and photons [109, 110]."

"The key challenge, in these cases, is represented by the engineering of the atom-photon coherent interactions to achieve strong cavity coupling and individual control of multiple resonant quantum emitters [105, 111, 112], that results to be affected by the decoherence introduced by the solid-state environment [113, 114]."

"A coherent multiple qubits interaction mediated via nano-cavity mode has been demonstrated in the microwave spectral region [115]."

"We numerically demonstrate that quantum emitters (QEs) can interact coherently with a sub-wavelength plasmonic nano-cavity by exploiting the overlap between the emitter near field and the gap plasmon mode."

Results and Discussion

"To control this resonant coupling we propose cascaded MIM designed to optimize the local field distribution and by tailoring narrow-banded spectral windows to in- and out-couple radiation from the nano-cavities."

"The plasmonic MIM cavity systems, if opportunely designed, can confine resonant epsilon-near-zero (ENZ) modes at one resonant wavelength that can be excited in absence of momentum matching techniques [116]."

"The top MIM structure is designed to support a canalization mode at 520 nm, finely tuned to out-couple the QEs radiated field."

"The MIM/QEs doped nano-cavity system is designed as a cascade to optimize the canalization of the QEs and as a consequence to enhance both the excitation and the emission."

"The field compression effect is a consequence of a spontaneous canalization of both the exciting and radiated field of QEs, occurring at the ENZ singularity point of the plasmonic MIM cavity."

Conclusion

"This work was aimed to demonstrate how the new degree of electromagnetic design freedom enabled by metamaterial nano-cavities can be combined with quantum emitters to: (i) enhance and modify microscopic light–matter interactions at the single emitter level to realize new regimes of solid-state quantum optics, and (ii) enable collective, coherent excitations in dense arrays of emitters to realize photonic analogues of quantum solids with new optoelectronic properties."

"At the quantum level, both the strength and nature of light-matter interactions are limited by the size mismatch between the optical wavelength and the electronic wavefunction of single emitters."

"We have demonstrated here, how manipulating the geometric and physical parameters of nano-cavity metamaterials can directly challenge this limit by shrinking or expanding the optical wavelength for a fixed operating frequency and can induce striking compression and canalization of the radiated field."

Nondestructive Classification of Quantum States Using an Algorithmic Quantum Computer [117]

This is a machine-generated summary of:

Babukhin, D. V.; Zhukov, A. A.; Pogosov, W. V.: Nondestructive classification of quantum states using an algorithmic quantum computer [117].

Published in: Quantum Machine Intelligence (2019).

Link to original: https://doi.org/10.1007/s42484-019-00010-9

If you want to cite the papers, please refer to the original.

For technical reasons we could not place the page where the original quote is coming from.

Abstract-Summary

"Methods of processing quantum data become more important as quantum computing devices improve their quality towards fault tolerant universal quantum computers."

"These methods include discrimination and filtering of quantum states given as an input to the device that may find numerous applications in quantum information technologies."

"This can be achieved by incorporating phase estimation algorithm into the hybrid quantum-classical computation scheme, where quantum block is trained classically."

"These ideas may find applications in other realization of hybrid quantum-classical computations with noisy quantum machines."

Introduction

"Quantum algorithms within such applications can be used as a part of a larger computation scheme which also incorporates classical blocks."

"Most of the proposals unfortunately are characterized by input/output bottlenecks occurring at stages of encoding classical data into quantum states and decoding them back (Aaronson [118]; Arunachalam and others [119])."

"The role of the quantum machine is to recognize their underlying patterns, which may have no classical counterpart (for example, characteristics of quantum entanglement), and then to classify these states or filter them."

"We address a hybrid quantum-classical approach to the problem of classification of input quantum states, where quantum block is trained classically with the set of labeled input vectors (supervised learning)."

"It is thus possible to make a classification of certain input quantum states both nondestructively and deterministically."

"This limitation restricts possible realizations of quantum machine learning algorithms to few-qubit examples, see, e.g., Cai and others [120] and Li and others [121]."

Phase Estimation Algorithm in Classification Problems

"If the input state is an eigenstate of $U(\omega)$, the measurements of ancilla qubits do not destroy it, so $|\psi\rangle$ is passed nondestructively through the scheme (apart of a general phase it obtains)."

"If there are two input states each being eigenstates of $U(\omega)$ with different eigenvalues of the above type, it is possible to classify these states both nondestructively and deterministically by doing measurements of ancillas."

"We may try to perform an ideal classification of these states, i.e., to construct an operator $U(\omega)$, for which these states are eigenstates and, moreover, the results of ancilla's measurements allow for the unambiguous deterministic discrimination between them."

"Let us stress that such a circuit provides a nondestructive and deterministic classification among given set of M input states, while a general input state is classified probabilistically."

"It is possible to find such a $U(\omega)$, which yields a nondestructive but probabilistic classification of M orthogonal training states."

Toy Model: Classification of Maximally Entangled Two-Qubit States

"We first show explicitly that such a parametrization for U gives a desirable result and also determine optimal values of ω_1 and ω_2 yielding nondestructive and deterministic classification."

"We see that indeed nondestructive and deterministic classification of two groups of input states is possible, since for $|\Phi_{\pm}\rangle$ the probability $P_0(|\Phi_{\pm}\rangle)$ to find ancilla in the state $|0\rangle$ is exactly 1, while for $|\Psi_{\pm}\rangle$ the probability $P_0(|\Psi_{\pm}\rangle)$ to find ancilla in the state $|0\rangle$ is exactly 0."

"For the input two-qubit state of a general form after some straightforward calculations, we obtain the expression for probability $P_0(|\Psi\rangle)$ to find ancilla in the state $|0\rangle$ provided optimal ω_1, $\omega_2 = 1$ are incorporated into the circuit It can be rewritten as In this general case, the scheme works as a probabilistic classifier, and the classification occurs according to the distance between the input state and two subspaces, in which $|\Phi_{\pm}\rangle$ and $|\Psi_{\pm}\rangle$ form local bases."

Implementation on a Noisy Quantum Device

"In order to use such devices for realization of quantum algorithms, one has to deal with the accumulation of errors."

"Classification of quantum states involving larger number of qubits implies application of much larger number of two-qubit gates which provide the main contribution to the total error rate."

"Automatic error correction or classical postselection of results can be then applied to discard part of wrong outputs associated with certain quantum errors."

"The uniformity of the wrong part of the output data with respect to this parameter was attributed to the fact that the circuit was not so shallow and contained a reasonable number of noisy quantum gates."

"Our procedure provides a case study which illustrates that it is possible to extract valuable information from data of noisy quantum computer even if they are heavily damaged by the decoherence and gate errors."

Conclusion

"We have addressed a hybrid quantum-classical scheme for the classification of input quantum states, where quantum part is represented by the phase estimation algorithm."

"For a general input quantum state, the scheme works as a probabilistic classifier and can be used to classify underlying patterns in quantum data."

"These ideas may be used in other realizations of hybrid quantum-classical computation schemes."

"Our results also demonstrate that pattern recognition can be an important ingredient of classical postprocessing of data from noisy quantum hardware."

References

1. Das S, Zhang J, Martina S, Suter D, Caruso F (2023) Quantum pattern recognition on real quantum processing units. Quantum Mach Intell. https://doi.org/10.1007/s42484-022-00093-x
2. Zhang Y, Lu K, Gao Y (2015) Qsobel: a novel quantum image edge extraction algorithm. Sci China Inf Sci 58(1):1–13. https://doi.org/10.1007/s11432-014-5158-9
3. Yao X-W, Wang H, Liao Z, Chen M-C, Pan J, Li J, Zhang K, Lin X, Wang Z, Luo Z, Zheng W, Li J, Zhao M, Peng X, Suter D (2017) Quantum image processing and its application to edge detection: theory and experiment. Phys Rev X 7:031041. https://doi.org/10.1103/PhysRevX.7.031041
4. Cavalieri G, Maio D (2020) A quantum edge detection algorithm. arXiv:2012.11036
5. Xu P, He Z, Qiu T, Ma H (2020) Quantum image processing algorithm using edge extraction based on kirsch operator. Opt Express 28(9):12508–12517. https://doi.org/10.1364/OE.386283
6. Neigovzen R, Neves JL, Sollacher R, Glaser SJ (2009) Quantum pattern recognition with liquid-state nuclear magnetic resonance. Phys Rev A 79:042321. https://doi.org/10.1103/PhysRevA.79.042321
7. Montanaro A (2015) Quantum pattern matching fast on average. https://doi.org/10.1007/s00453-015-0060-4
8. Zhang Y, Lu K, Xu K, Gao Y, Wilson RC (2015) Local feature point extraction for quantum images. Quantum Inf Process 14:1573–1588
9. Jiang N, Dang Y, Wang J (2016) Quantum image matching. Quantum Inf Process 15(9):3543–3572. https://doi.org/10.1007/s11128-016-1364-2
10. Soni KK, Rasool A (2020) Pattern matching: a quantum oriented approach. Procedia Comput Sci 167:1991–2002. https://doi.org/10.1016/j.procs.2020.03.230. International Conference on Computational Intelligence and Data Science
11. Tezuka H, Nakaji K, Satoh T, Yamamoto N (2022) Grover search revisited: application to image pattern matching. Phys Rev A 105:032440. https://doi.org/10.1103/PhysRevA.105.032440
12. Jozsa R (1994) Fidelity for mixed quantum states. J Mod Opt 41(12):2315–2323
13. Schumacher B (1995) Quantum coding. Phys Rev A 51(4):2738
14. Garcia-Escartin JC, Chamorro-Posada P (2013) Swap test and hong-ou-mandel effect are equivalent. Phys Rev A 87:052330. https://doi.org/10.1103/PhysRevA.87.052330
15. Wrachtrup J, Jelezko F (2006) Processing quantum information in diamond. J Phys Condens Matter 18:807–824. https://doi.org/10.1088/0953-8984/18/21/S08
16. Suter D, Jelezko F (2017) Single-spin magnetic resonance in the nitrogen-vacancy center of diamond. Prog Nucl Magn Reson Spectrosc 98-99:50–62. https://doi.org/10.1016/j.pnmrs.2016.12.001
17. Di S, Xu J, Shu G, Feng C, Ding X, Shan Z (2023) Amplitude transformed quantum convolutional neural network. Appl Intell. https://doi.org/10.1007/s10489-023-04581-w
18. Cong I, Choi S, Lukin MD (2019) Quantum convolutional neural networks. Nat Phys 15(12):1273–1278
19. Broughton M, Verdon G, McCourt T, Martinez AJ, Yoo JH, Isakov SV, Massey P, Halavati R, Niu MY, Zlokapa A et al (2020) Tensorflow quantum: a software framework for quantum machine learning. arXiv:2003.02989
20. Sim S, Johnson PD, Aspuru-Guzik A (2019) Expressibility and entangling capability of parameterized quantum circuits for hybrid quantum-classical algorithms. Adv Quantum Technol 2(12):1900070
21. Henderson M, Shakya S, Pradhan S, Cook T (2020) Quanvolutional neural networks: powering image recognition with quantum circuits. Quantum Mach Intell 2(1):2
22. Bastrakova MV, Kulandin DS, Laptyeva T, Vozhakov VA, Liniov AV (2023) Genetic algorithm for searching bipolar single-flux-quantum pulse sequences for qubit control. Lobachevskii J Math. https://doi.org/10.1134/s1995080223010043

23. McDermott R, Vavilov M (2014) Accurate qubit control with single flux quantum pulses. Phys Rev Appl 2:014007
24. Li K, McDermott R, Vavilov MG (2019) Hardware-efficient qubit control with single-flux-quantum pulse sequences. Phys Rev Appl 12(1):014044
25. Koch J, Terri MY, Gambetta J, Houck AA, Schuster DI, Majer J, Blais A, Devoret MH, Girvin SM, Schoelkopf RJ (2007) Charge-insensitive qubit design derived from the Cooper pair box. Phys Rev A 76:042319
26. Chow JM, Corcoles AD, Gambetta JM, Rigetti C, Johnson BR, Smolin JA, Rozen JR, Keefe GA, Rothwell MB, Ketchen MB, Steffen M (2011) Simple all-microwave entangling gate for fixed-frequency superconducting qubits. Phys Rev Lett 107:080502
27. Ram J, Dutt D, Dhiman SK, Behera BK, Panigrahi PK (2023) Entanglement concentration of multi-qubit entangled states: an IBM quantum experience. Quantum Stud Math Foundations. https://doi.org/10.1007/s40509-023-00298-0
28. Satyajit S, Srinivasan K, Bahera BK, Panigarhi PK (2018) Nondestructive discrimination of a new family of highly entangled states in IBM quantum computer. Quant Inf Process 17:212
29. Briegel HJ, Raussendorf R (2001) Persistent entanglement in arrays of interacting particles. Phys Rev Lett 86:910
30. Raussendorf R, Briegel HJ (2001) A one-way quantum computer. Phys Rev Lett 86:5188
31. Hein M, Dur W, Eisert J, Raussendorf R, Van den Nest M, Briegel H-J. Entanglement in graph states and its applications. arXiv:quant-ph/0602096
32. Cervera-Lierta A, Lattore JI, Goyeneche D (2019) Quantum circuits for maximally entangled states. Phys Rev A 100:022342
33. Preskill J (2018) Quantum computing in the NISQ era and beyond. Quantum 2:79. https://doi.org/10.22331/q-2018-08-06-79
34. Peruzzo A, McClean J, Shadbolt P, Yung MH, Zhou XQ, Love PJ, Aspuru-Guzik A, O'brien JL (2014) A variational eigenvalue solver on a photonic quantum processor. Nat Commun 5:4213. https://doi.org/10.1038/ncomms5213
35. Niu J, Zhang L, Liu Y, Qiu J, Huang W, Huang J, Jia H, Liu J, Tao Z, Wei W, Zhou Y, Zou W, Chen Y, Deng X, Deng X, Hu C, Hu L, Li J, Tan D, Xu Y, Yan F, Yan T, Liu S, Zhong Y, Cleland AN, Yu D (2023) Low-loss interconnects for modular superconducting quantum processors. Nat Electron. https://doi.org/10.1038/s41928-023-00925-z
36. Ball P (2021) First quantum computer to pack 100 qubits enters crowded race. Nature 599:542
37. Burkhart LD et al (2021) Error-detected state transfer and entanglement in a superconducting quantum network. PRX Quantum 2:030321
38. Awschalom D et al (2021) Development of quantum interconnects (QuICs) for next-generation information technologies. PRX Quantum 2:017002
39. Kurpiers P et al (2018) Deterministic quantum state transfer and remote entanglement using microwave photons. Nature 558:264–267
40. Axline CJ et al (2018) On-demand quantum state transfer and entanglement between remote microwave cavity memories. Nat Phys 14:705–710
41. Campagne-Ibarcq P et al (2018) Deterministic remote entanglement of superconducting circuits through microwave twophoton transitions. Phys Rev Lett 120:200501
42. Leung N et al (2019) Deterministic bidirectional communication and remote entanglement generation between superconducting qubits. Quantum Inf 5:18
43. Magnard P et al (2020) Microwave quantum link between superconducting circuits housed in spatially separated cryogenic systems. Phys Rev Lett 125:260502
44. Zhong Y et al (2021) Deterministic multi-qubit entanglement in a quantum network. Nature 590:571–575
45. Wang Y-D, Clerk AA (2012) Using dark modes for high-fidelity optomechanical quantum state transfer. New J Phys 14:105010
46. Greenberger DM, Horne MA, Shimony A, Zeilinger A (1990) Bell's theorem without inequalities. Am J Phys 58:1131–1143

47. Song C et al (2019) Generation of multicomponent atomic Schrödinger cat states of up to 20 qubits. Science 365:574–577
48. Meichanetzidis K, Toumi A, de Felice G, Coecke B (2023) Grammar-aware sentence classification on quantum computers. Quantum. Mach Intell. https://doi.org/10.1007/s42484-023-00097-1
49. Coecke B, Sadrzadeh M, Clark S (2010) Mathematical foundations for a compositional distributional model of meaning. https://arxiv.org/abs/1003.4394
50. Grefenstette E, Sadrzadeh M (2011) Experimental support for a categorical compositional distributional model of meaning. In: The 2014 conference on empirical methods on natural language processing. arXiv.: 1106.4058, pp 1394–1404
51. Kartsaklis D, Sadrzadeh M (2013) Prior disambiguation of word tensors for constructing sentence vectors. In: The 2013 conference on empirical methods on natural language processing. ACL, pp 1590–1601
52. Zeng W, Coecke B (2016) Quantum algorithms for compositional natural language processing. Electr Proc Theor Comput Sci 221:67–75. https://doi.org/10.4204/eptcs.221.8
53. O'Riordan LJ, Doyle M, Baruffa F, Kannan V (2020) A hybrid classical-quantum workflow for natural language processing. Mach Learn Sci Technol. https://doi.org/10.1088/2632-2153/abbd2e
54. Wiebe N, Bocharov A, Smolensky P, Troyer M, Svore KM (2019) Quantum Language Processing. https://arxiv.org/abs/1902.05162
55. Bausch J, Subramanian S, Piddock S (2020) A quantum search decoder for natural language processing. https://arxiv.org/abs/1909.05023
56. Chen JC (2002) Quantum computation and natural language processing
57. Abramsky S, Coecke B (2004) A categorical semantics of quantum protocols. In: Proceedings of the 19th annual IEEE symposium on logic in computer science, 2004, pp 415–425. https://doi.org/10.1109/LICS.2004.1319636
58. Meichanetzidis K, Gogioso S, Felice GD, Chiappori N, Toumi A, Coecke B (2020) Quantum natural language processing on near-term quantum computers. https://arxiv.org/abs/2005.04147
59. de Felice G, Meichanetzidis K, Toumi A (2020) Functorial question answering. Electron Proc Theor Comput Sci 323:84–94. https://doi.org/10.4204/eptcs.323.6
60. Chen Y, Pan Y, Dong D (2020) Quantum language model with entanglement embedding for question answering. https://arxiv.org/abs/2008.09943
61. Zhao Q, Hou C, Liu C, Zhang P, Xu R (2020) A quantum expectation value based language model with application to question answering. Entropy 22(5):533
62. Coecke B (2020) The mathematics of text structure. https://arxiv.org/abs/1904.03478
63. Coecke B, de Felice G, Meichanetzidis K, Toumi A (2020) Foundations for near-term quantum natural language processing. https://arxiv.org/abs/2012.03755
64. Ren W, Li W, Xu S, Wang K, Jiang W, Jin F, Zhu X, Chen J, Song Z, Zhang P, Dong H, Zhang X, Deng J, Gao Y, Zhang C, Wu Y, Zhang B, Guo Q, Li H, Wang Z, Biamonte J, Song C, Deng D-L, Wang H (2022) Experimental quantum adversarial learning with programmable superconducting qubits. Nat Comput Sci. https://doi.org/10.1038/s43588-022-00351-9
65. Lu S, Duan L-M, Deng D-L (2020) Quantum adversarial machine learning. Phys Rev Res 2:033212
66. Li W, Deng D-L (2022) Recent advances for quantum classifiers. Sci China Phys Mech Astron 65:220301
67. Havlíček V, Córcoles AD, Temme K, Harrow AW, Kandala A, Chow JM, Gambetta JM (2019) Supervised learning with quantum-enhanced feature spaces. Nature 567(7747):209–212
68. Gong M et al. (2022) Quantum neuronal sensing of quantum many-body states on a 61-qubit programmable superconducting processor. Preprint at https://arxiv.org/abs/2201.05957
69. Herrmann J, Llima SM, Remm A, Zapletal P, McMahon NA, Scarato C, Swiadek CK, Hellings C, Krinner S et al (2022) Realizing quantum convolutional neural networks on a superconducting quantum processor to recognize quantum phases. Nat Commun 13(1):4144

70. Bennewitz ER, Hopfmueller F, Kulchytskyy B, Carrasquilla J, Ronagh P (2022) Neural error mitigation of near-term quantum simulations. Nat Mach Intell. https://doi.org/10.1038/s42256-022-00509-0
71. Huembeli P, Dauphin A (2021) Characterizing the loss landscape of variational quantum circuits. Quantum Sci Technol 6:025011
72. Park C-Y, Kastoryano MJ (2020) Geometry of learning neural quantum states. Phys Rev Res 2:023232
73. Bukov M, Schmitt M, Dupont M (2021) Learning the ground state of a non-stoquastic quantum Hamiltonian in a rugged neural network landscape. Sci Post Phys 10:147
74. Qiskit Development Team. Qiskit: an open source framework for quantum computation, ver. 0.23.0. https://qiskit.org (IBM, 2019)
75. LaRose R, Rieffel E, Venturelli D (2022) Mixer-phaser Ansätze for quantum optimization with hard constraints. Quantum Mach Intell. https://doi.org/10.1007/s42484-022-00069-x
76. Hadfield S, Wang Z, O'Gorman B, Rieffel EG, Venturelli D, Biswas R (2019) From the quantum approximate optimization algorithm to a quantum alternating operator ansatz. Algorithms 12(2):34
77. Do M, Wang Z, O'Gorman B, Venturelli D, Rieffel E, Frank J (2020) Planning for compilation of a quantum algorithm for graph coloring. In: Proceedings of the 24th European conference on artificial intelligence (ECAI'2020)
78. Wang Z, Rubin NC, Dominy JM, Rieffel EG (2020) XY mixers: analytical and numerical results for the quantum alternating operator ansatz. Phys Rev A 101(1):012320
79. Egger DJ, Mareček J, Woerner S (2021) Warm-starting quantum optimization. Quantum 5:479
80. Kim M, Venturelli D, Jamieson K (2019) Leveraging quantum annealing for large mimo processing in centralized radio access networks. In: Proceedings of the ACM special interest group on data communication, pp 241–255
81. Streif M, Leib M, Wudarski F, Rieffel E, Wang Z (2021) Quantum algorithms with local particle-number conservation: noise effects and error correction. Phys Rev A 103(4):042412
82. Hashim A, Rines R, Omole V, Naik RK, Kreikebaum JM, Santiago DI, Chong FT, Siddiqi I, Gokhale P (2021) Optimized fermionic swap networks with equivalent circuit averaging for qaoa. arXiv:2111.04572
83. Zhao C, Gao X-S (2021) QDNN: deep neural networks with quantum layers. Quantum. Mach Intell. https://doi.org/10.1007/s42484-021-00046-w
84. Killoran N, Bromley TR, Arrazola JM, Schuld M, Quesada N, Lloyd S (2019) Continuous-variable quantum neural networks. Phys Rev Res 1(3):033,063
85. Zhao J, Zhang YH, Shao CP, Wu YC, Guo GC, Guo GP (2019) Building quantum neural networks based on a swap test. Phys Rev A 100(012):334. https://doi.org/10.1103/PhysRevA.100.012334
86. Kerenidis I, Landman J, Prakash A (2020) Quantum algorithms for deep convolutional neural networks. In: International conference on learning representations. https://openreview.net/forum?id=Hygab1rKDS
87. Lloyd S, Weedbrook C (2018) Quantum generative adversarial learning. Phys Rev Lett 121(4):040,502
88. Dallaire-Demers PL, Killoran N (2018) Quantum generative adversarial networks. Phys Rev A 98(1):012,324
89. Liu JG, Wang L (2018) Differentiable learning of quantum circuit born machines. Phys Rev A 98(6):062,324
90. Schuld M, Killoran N (2019) Quantum machine learning in feature hilbert spaces. Phys Rev Lett 122(4):040,504
91. Benedetti M, Lloyd E, Sack S, Fiorentini M (2019) Parameterized quantum circuits as machine learning models. Quantum. Sci Technol 4(4):043,001
92. McClean JR, Romero J, Babbush R, Aspuru-Guzik A (2016) The theory of variational hybrid quantum-classical algorithms. New J Phys 18(2):023,023

93. Liu JG, Zhang YH, Wan Y, Wang L (2019) Variational quantum eigensolver with fewer qubits. Phys Rev Res 1(023):025. https://doi.org/10.1103/PhysRevResearch.1.023025
94. Cybenko G (1989) Approximation by superpositions of a sigmoidal function. Math Control Signals Syst 2(4):303–314
95. Hornik K (1991) Approximation capabilities of multilayer feedforward networks. Neural Netw 4(2):251–257. https://doi.org/10.1016/0893-6080(91)90009-T. http://www.sciencedirect.com/science/article/pii/089360809190009T
96. Leshno M, Lin VY, Pinkus A, Schocken S (1993) Multilayer feedforward networks with a nonpolynomial activation function can approximate any function. Neural Netw 6(6):861–867. https://doi.org/10.1016/S0893-6080(05)80131-5. http://www.sciencedirect.com/science/article/pii/S0893608005801315
97. Pinkus A (1999) Approximation theory of the mlp model in neural networks. Acta Numerica 8:143–195. https://doi.org/10.1017/S0962492900002919
98. Cerezo M, Sone A, Volkoff T, Cincio L, Coles PJ (2020) Cost-function-dependent barren plateaus in shallow quantum neural networks. arXiv: 2001.00550
99. Kingma DP, Ba J (2014) Adam: a method for stochastic optimization. arXiv: 1412.6980
100. Huang C, Zhang F, Newman M, Ni X, Ding D, Cai J, Gao X, Wang T, Wu F, Zhang G, Ku H-S, Tian Z, Wu J, Xu H, Yu H, Yuan B, Szegedy M, Shi Y, Zhao H-H, Deng C, Chen J (2021) Efficient parallelization of tensor network contraction for simulating quantum computation. Nat Comput Sci. https://doi.org/10.1038/s43588-021-00119-7
101. Biamonte JD, Morton J, Turner J (2015) Tensor network contractions for #SAT. J Stat Phys 160:1389–1404
102. Huang C, Newman M, Szegedy M (2020) Explicit lower bounds on strong quantum simulation. IEEE Trans Inf Theory 66:5585–5600
103. Palermo G, Lio GE, Strangi G (2020) Compressed and canalized emission of quantum emitters in MIM nano-cavities. Quantum Stud Math Found. https://doi.org/10.1007/s40509-020-00231-9
104. Faraon A, Majumdar A, Englund D, Kim E, Bajcsy M, Vučković J (2011) Integrated quantum optical networks based on quantum dots and photonic crystals. New J Phys 13(5):055025
105. Sipahigil A, Evans RE, Sukachev DD, Burek MJ, Borregaard J, Bhaskar MK, Nguyen CT, Pacheco JL, Atikian HA, Meuwly C et al (2016) An integrated diamond nanophotonics platform for quantum-optical networks. Science 354(6314):847
106. Stannigel K, Rabl P, Zoller P (2012) Driven-dissipative preparation of entangled states in cascaded quantum-optical networks. New J Phys 14(6):063014
107. DiCarlo L et al (2009) Demonstration of two-qubit algorithms with a superconducting quantum processor. Nature 460:240–244
108. Fu X, Rol MA, Bultink CC, Van Someren J, Khammassi N, Ashraf I, Vermeulen R, De Sterke J, Vlothuizen W, Schouten R et al. (2017). An experimental microarchitecture for a superconducting quantum processor. In: Proceedings of the 50th Annual IEEE/ACM International Symposium on Microarchitecture, pp. 813–825
109. Northup T, Blatt R (2014) Quantum information transfer using photons. Nat Photonics 8(5):356
110. Zhang J, Liu YX, Li CW, Tarn TJ, Nori F (2009) Generating stationary entangled states in superconducting qubits. Phys Rev A 79(5):052308
111. Laucht A, Villas-Bôas J, Stobbe S, Hauke N, Hofbauer F, Böhm G, Lodahl P, Amann MC, Kaniber M, Finley J (2010) Mutual coupling of two semiconductor quantum dots via an optical nanocavity. Phys Rev B 82(7):075305
112. Lio GE, Madrigal JB, Xu X, Peng Y, Pierini S, Couteau C, Jradi S, Bachelot R, Caputo R, Blaize S (2019) Integration of nanoemitters onto photonic structures by guided evanescent-wave nano-photopolymerization. J Phys Chem C 123(23):14669
113. Lodahl P, Mahmoodian S, Stobbe S (2015) Interfacing single photons and single quantum dots with photonic nanostructures. Rev Mod Phys 87(2):347

114. Kim H, Sridharan D, Shen TC, Solomon GS, Waks E (2011) Strong coupling between two quantum dots and a photonic crystal cavity using magnetic field tuning. Opt Express 19(3):2589
115. Majer J, Chow J, Gambetta J, Koch J, Johnson B, Schreier J, Frunzio L, Schuster D, Houck AA, Wallraff A et al (2007) Coupling superconducting qubits via a cavity bus. Nature 449(7161):443
116. Passler NC, Gubbin CR, Folland TG, Razdolski I, Katzer DS, Storm DF, Wolf M, De Liberato S, Caldwell JD, Paarmann A (2018) Strong coupling of epsilon-near-zero phonon polaritons in polar dielectric heterostructures. Nano Lett 18(7):4285
117. Babukhin DV, Zhukov AA, Pogosov WV (2019) Nondestructive classification of quantum states using an algorithmic quantum computer. Quantum Mach Intell. https://doi.org/10.1007/s42484-019-00010-9
118. Aaronson S (2015) Read the fine print. Nat Phys 11:291–293. https://doi.org/10.1038/nphys3272
119. Arunachalam S, Gheorghiu V, Jochym-O'Connor T, Mosca M, Srinivasan PV (2015) On the robustness of bucket brigade quantum RAM. New J Phys 17:123010
120. Cai X-D, Wu D, Su Z-E, Chen M-C, Wang X-L, Li L, Liu N-L, Lu C-Y, Pan J-W (2015) Entanglement-based machine learning on a quantum computer. Phys Rev Lett 114:110504
121. Zhaokai L, Xiaomei L, Nanyang X, Jiangfeng D (2015) Experimental realization of a quantum support vector machine. Phys Rev Lett 114:140504. https://doi.org/10.1103/PhysRevLett.114.140504

Chapter 5
Quantum Machine Learning (QML)

Introduction by the Editor

This chapter deals with the study of quantum machine learning which aims to achieve a speedup over traditional machine learning for data analysis. Further, deciphering tensor network quantum machine learning models, variational quantum circuits for quantum machine learning models, near-optimal quantum algorithms for string problems, quantum neural network for time-series predictions, Information-theoretically secure quantum encryption, quantum methods of temporal space optimization for real-time IoT applications, network attack detection scheme based on variational quantum neural network and quantum convolutional neural network for data analysis have been highlighted.

In this chapter, we address the trainability challenges in quantum machine learning (QML), a field that aims to leverage quantum computational devices and quantum models for data analysis tasks. While QML holds the potential for significant speedups over traditional machine learning approaches, efficient training of the parameters in quantum neural networks remains a key challenge. We note that prior theoretical scaling results for trainability have been established in the closely related field of variational quantum algorithms (VQAs). However, the differences between VQAs and QML models pose challenges or technical issues in directly applying these results to the latter. In the research study, we bridge the two frameworks and demonstrate that gradient scaling results for VQAs can also be applied to analyze the trainability of QML models.

In this chapter, we investigate the applicability of variational quantum circuits as parameterized policies in reinforcement learning tasks. Specifically, we employ a low-depth hardware-efficient ansatz for the variational quantum circuit, which serves as the policy of a reinforcement learning agent. The research study focuses on understanding the effectiveness of quantum models in reinforcement learning settings and exploring their advantages compared to classical neural networks.

V. Natarajan (ed.), *Quantum Artificial Intelligence*,
https://doi.org/10.1007/978-981-96-5051-4_5

Furthermore, we investigate the phenomenon of barren plateaus in quantum policy gradients by analyzing the spectrum of the Fisher information matrix. By studying the characteristics of barren plateaus in the context of quantum reinforcement learning, we gain insights into the challenges and limitations of training variational quantum circuits for policy optimization.

In this chapter, we bring out "Near-Optimal Quantum Algorithms for String Problems" that describes quantum algorithms for fundamental string problems such as Longest Common Substring, Lexicographically Minimal String Rotation, and Longest Square Substring. These algorithms solve near-optimal query and time complexities. We also discuss computational complexity of the Longest Common Substring problem and MNRS quantum walk framework. Next, we solve three important problems in quantum algorithms (Near-field classical work): (1) Longest common substring, (2) Minimal string rotation and (3) Longest square substring.

The research article "A Quantum-Inspired Online Spiking Neural Network for Time-Series Predictions" describes the use of quantum-inspired techniques to enhance spiking neural networks (SNNs) for time-series prediction tasks. SNNs are well known for their energy efficiency and dynamic structure, which mimics biological brains. The research study aims at online learning, a method suitable for dynamic and continuously changing environments.

An innovative model utilizes quantum-inspired optimization algorithms to improve the training and performance of SNNs. Specifically, the research employs a quantum particle swarm optimization (QPSO) algorithm to optimize the network parameters effectively. The outcome model is tested on various time-series prediction tasks, demonstrating superior performance compared to traditional methods.

In the realm of data processing and IoT communication, the fusion of quantum algorithms with metaheuristics presents an avenue for developing more robust solutions with potential applications across various industries. These hybrid algorithms aim to optimize various aspects such as complexity, efficiency, processing speed, and accuracy, thereby addressing real-world challenges, including reducing occupational risks in communication systems.

In QML work, we explore quantum algorithms for fundamental string problems, including Longest Common Substring, Lexicographically Minimal String Rotation, and Longest Square Substring. While these problems have been extensively studied in classical stringology literature since the 1970s and have near-linear time classical algorithms, we aim to provide quantum algorithms with near-optimal query and time complexities.

In this chapter, we propose a novel approach called Quantum-Inspired Online Spiking Neural Network (QiSNN) designed to address the challenges faced by traditional spiking neural networks (SNNs) in dynamic scenarios involving temporal sequences. The motivation behind QiSNN stems from the superior dynamic structures and low energy consumption exhibited by SNNs, resembling those of the biological brain.

In this chapter we mention an industrial application of QC. Integrating quantum procedures into real industrial pipelines, especially using data from physical machines like those in Eni's Oil Treatment Plants, represents a significant step forward in the practical application of quantum computing. Next the focus is on quantum encryption schemes with imperfect secrecy and imperfect correctness, providing a systematic study of information-theoretically secure quantum encryption. The quantum research explores two secrecy definitions: One weaker and One stronger. More details of them are explained in this chapter. Further, quantum research presents a novel approach to optimizing IoT data acquisition for real-time applications using techniques inspired by quantum computing principles. The goal is to improve accuracy and efficiency in time-sensitive data analysis and decision-making processes. The study quantifies IoT sensors in terms of Sensors of Interest (SoI) and Degree of Aptness (DoA) measure.

Quantum research presents a novel approach to addressing the learning parity with noise (LPN) problem, which involves learning a hidden parity function from noisy data. LPN is considered an example of intelligent behavior as it involves generalizing a concept based on noisy examples. The solution to LPN is also relevant to decoding random binary linear codes in the presence of classification noise. While solving LPN efficiently is thought to be intractable classically, it can be achieved efficiently using a quantum oracle. Further, we propose an intrusion detection scheme based on a variational quantum neural network (VQNN), which combines a variational quantum circuit (VQC) with classical machine learning (ML) strategies. The goal is to leverage the quantum advantage demonstrated by VQNN in classification problems to improve the accuracy and efficiency of intrusion detection systems (IDS) compared to traditional machine learning-based IDS. The research work "Network Attack Detection Scheme Based on Variational Quantum Neural Network" describes an innovative method for detecting network attacks using variational quantum neural networks (VQNNs). This method leverages the advantages of quantum machine learning, which can offer higher accuracy and efficiency compared to traditional machine learning methods for intrusion detection systems (IDS). Hence VQNNs are efficient and accurate detection of networks, identify complex patterns and improving cybersecurity measures.

At the end of Chap. 5, we discuss the QCNN model which is inspired by convolutional neural networks (CNNs) commonly used in classical machine learning but operates on quantum circuits with only two-qubit interactions. The suitability of QCNN with NISQ devices are described here. QCNNs serves for encoding classical data into quantum states, including amplitude encoding, qubit encoding, dense qubit encoding, and hybrid encoding, each with its trade-offs regarding quantum circuit depth and width.

Machine Generated Summaries

Disclaimer: The summaries in this chapter were generated from Springer Nature publications using extractive AI auto-summarization: An extraction-based summarizer aims to identify the most important sentences of a text using an algorithm and uses those original sentences to create the auto-summary (unlike generative AI). As the constituted sentences are machine selected, they may not fully reflect the body of the work, so we strongly advise that the original content is read and cited. The auto generated summaries were curated by the editor to meet Springer Nature publication standards. To cite this content, please refer to the original papers.

Machine generated keywords: learning, neural, qml, network, classical datum, neural network, iot, string, quantum neural, machine, qnn, machine learn, gradient, ancilla, model.

Subtleties in the Trainability of Quantum Machine Learning Models [1]

This is a machine-generated summary of:

Thanasilp, Supanut; Wang, Samson; Nghiem, Nhat Anh; Coles, Patrick; Cerezo, Marco: Subtleties in the trainability of quantum machine learning models [1].

Published in: Quantum Machine Intelligence (2023).

Link to original: https://doi.org/10.1007/s42484-023-00103-6

If you want to cite the papers, please refer to the original.

For technical reasons we could not place the page where the original quote is coming from.

Abstract-Summary

"This field, called quantum machine learning (QML), aims to achieve a speedup over traditional machine learning for data analysis."

"Its success usually hinges on efficiently training the parameters in quantum neural networks, and the field of QML is still lacking theoretical scaling results for their trainability."

"Some trainability results have been proven for a closely related field called variational quantum algorithms (VQAs)."

"We bridge the two frameworks and show that gradient scaling results for VQAs can also be applied to study the gradient scaling of QML models."

"Our results indicate that features deemed detrimental for VQA trainability can also lead to issues such as barren plateaus in QML."

"We provide theoretical and numerical evidence that QML models exhibit further trainability issues not present in VQAs, arising from the use of a training dataset."

Introduction

"We study the trainability and the existence of barren plateaus in QML models."

"Our work represents a general treatment that goes beyond previous analysis of gradient scaling and trainability in specific QML models (Pesah and others [2]; Sharma and others [3]; Liu and others [4]; Abbas and others [5]; Haug and Kim [6]; Kieferova and others [7]; Kiani and others [8]; Tangpanitanon and others [9])."

"We rigorously connect the scaling of gradients in VQA-type cost functions and QML loss functions, so that barren plateau results in the variational algorithms literature can be used to study the trainability of QML models."

"These results show that additional care must be taken when studying the trainability of QML models."

"We note that, while this work focuses on how barren plateaus affect the trainability of QNNs, a closely-related phenomena known as exponential concentration has also been studied in quantum kernel-based models, which another popular approach of QML (Thanasilp and others [10])."

Framework

"For the QML model to access the exponentially large dimension of the Hilbert space, one needs to encode the classical data into a quantum state."

"Since BPs have been mainly studied in the context of variational quantum algorithms (VQAs) (Cerezo and others [11]; McClean and others [12]; Cerezo and others [13]; Larocca and others [14]; Marrero and others [15]; Patti and others [16]; Holmes and others [17]; Holmes and others [18]; Huembeli and Dauphin [19]; Zhao and Gao [20]; Wang and others [21]), we here briefly recall that in a VQA implementation, the goal is to minimize a linear cost function that is usually of the form Here ρ is the initial state, $U(\theta)$ a trainable parametrized quantum circuit, and H a Hermitian operator."

"Let us first recall that, as previously mentioned, training a VQA or the QNN in a QML model usually implies optimizing the parameters in a quantum circuit."

"While VQAs generally deal with quantum states, QML models are envisioned to work both on classical and quantum data."

"When the QML model deals with quantum data, the input states are usually non-trivial states."

Analytical Results

"Corollary 1 explicitly implies that, under mild assumptions, previously established BP results for VQAs are applicable to QML models that utilize mean square error and negative log-likelihood loss functions."

"We argue that the dataset can negatively impact QML trainability if the input states to the QNN have high levels of entanglement and if the QNN employs local gates (which is the standard assumption in most widely used QNNs Farhi and Neven [22]; Cong and others [23]; Beer and others [24]; Bausch [25])."

"To illustrate this issue, we will present an example where a VQA does not exhibit a BP, but a QML model can still have a dataset-induced BP."

"For a QML setting, Proposition 1, combined with Theorem 1, implies that the QML model is susceptible to dataset-induced BPs even with simple QNNs and local measurements."

"Previously established BP results that hold independently of the input state such as global cost functions (Cerezo and others [13]), deep circuits (McClean and others [12]), expressibility (Holmes and others [17]), and noise (Wang and others [21]) will hold regardless of the embedding strategy."

Numerical Results

"Our results here show that even for a trivial QNN, and independently of the dataset and embedding scheme, global measurements in the loss function lead to exponentially small gradients, and thus to BPs in QML models."

"Since both the tensor product QNN with local cost and the QCNN are not expected to exhibit BPs with no training data and separable input states (see Cerezo and others [13] and Pesah and others [2], respectively), any unfavorable scaling arising here will be due to the structure of the data or the embedding scheme."

"One can see here that, independently of the structure of the dataset, the large entangling power of the embedding scheme leads to states that are essentially maximally mixed on any reduced pair of qubits."

"QNNs that have no BPs when trained on trivial input states can have exponentially vanishing gradients arising from either the structure of the dataset, or the large entangling power of the embedding scheme."

Implications for the Literature

"This is particularly relevant to the use of global observables such as measuring the parity of the output bitstrings on all qubits, which have been employed in the QML literature."

"In our numerics section, local parity measurements are practically useful and one can use them to optimize the model and achieve small training and generalization errors."

"As proven here, the choice of embedding cannot practice and mitigate the effect of BPs or prevent a BP that would otherwise exist for the QNN."

"The embedding cannot prevent a BP arising from the use of a global measurement, or from the use of a QNN that forms a 2-design."

"While embeddings can lead to a novel source of BPs, they cannot cure a BP that a particular QNN suffers from."

Discussion

"While this novel, general paradigm for data analysis is exciting, there are still very few theoretical results studying the scalability of QML."

"While VQAs and QML models share some similarities in that both train parametrized quantum circuits, there are some key differences that make it difficult to directly apply VQA trainability results to the QML setting."

"We bridged the gap between the VQA and QML frameworks by rigorously showing that gradient scaling results from VQAs will hold in a QML setting."

"Our results illustrate another subtlety that arises when using classical data, as the classically-hard-to-simulate embedding of Havlíček and others [26] leads to large generalization error on a standard MNIST classification task."

"Our results illuminate some subtleties in training QNNs in QML models, and show that more work needs to be done to guarantee that QML schemes will be trainable, and thus useful for practical applications."

Decohering Tensor Network Quantum Machine Learning Models [27]

This is a machine-generated summary of:

Liao, Haoran; Convy, Ian; Yang, Zhibo; Whaley, K. Birgitta: Decohering tensor network quantum machine learning models [27].

Published in: Quantum Machine Intelligence (2023).

Link to original: https://doi.org/10.1007/s42484-022-00095-9

If you want to cite the papers, please refer to the original.

For technical reasons we could not place the page where the original quote is coming from.

Abstract-Summary

"While decoherence of qubits is expected to decrease the performance of QML models, it is unclear to what extent the diminished performance can be compensated for by adding ancillas to the models and accordingly increasing the virtual bond dimension of the models."

"We investigate here the competition between decoherence and adding ancillas on the classification performance of two models, with an analysis of the decoherence effect from the perspective of regression."

Introduction

"To applying TNs such as the tree tensor network (TTN) (Shi and others [28]) and the multiscale entanglement renormalization ansatz (MERA) (Vidal [29]) for quantum-inspired tensor network ML algorithms (Stoudenmire [30]; Reyes and Stoudenmire [31]; Wall and D'Aguanno [32]), there have been efforts to variationally train the generic unitary nodes in TNs to perform quantum machine learning (QML) on data-encoded qubits."

"We investigate and numerically quantify the competing effect between decoherence and increasing bond dimension of two common tensor network QML models, namely the unitary TTN and the MERA."

"As we add and increase the number of ancillas and accordingly increase the virtual bond dimension of the tensor networks, this diminished expressiveness may be compensated by the increased dimension of the classical probability distributions and their conditionals, manifested in the increasing number of diagonals intermediate within the network, as well as by the increased sized of the stochastic matrices encapsulated by the corresponding Bayesian networks in the fully dephased limit."

Preliminaries

"Every node in a unitary TTN is forced to be a unitary matrix with respect to its input and output Hilbert spaces."

"If the classification is binary, at the output of the last layer, namely the root node, only one qubit is measured."

"After variationally learning the weights in the unitary nodes, we recover a quantum channel such that the information contained in the output qubits of each layer can be viewed as a coarse-grained representation of that in the input qubits, which sequentially extracts useful features of the data encoded in the data qubits."

"Operationally an isometry is replaced by a unitary node, half of whose output qubits are partially traced over, which is the same as a unitary node in the TTN."

Dephasing

"The suppression is stronger by a factor of $(1 - p)^2$ for regressors that are anti-diagonals of the input density matrix, i.e., ρ_{30} and ρ_{21}."

"Viewing the regressors at the input of the last layer, the suppression on most of them by some power of $(1 - p)$ resembles the concept of regularization in regressions but does not involve a penalty term on the coefficient norm in the loss function."

"In cases where there can be entanglement in each of the input qubits, such as the intermediate layers in a MERA or in a unitary TTN with ancillas, the pattern of suppressing certain regressors is similar, where the coherence of the input is suppressed by some power of $(1 - p)$."

"The regressors on the anti-diagonals are most strongly suppressed by a factor of $(1 - p)^m$ where m is the number of input qubits."

Adding Ancillas and Increasing the Virtual Bond Dimension

"In the scheme of adding ancillas per node in a unitary TTN, every node requires then in principle at least two ancilla qubits to achieve an arbitrary quantum channel, because there are two input qubits coming from the previous layer and one output qubit passing to the next layer."

"Although the ancilla-per-data-qubit scheme achieves superior classification performance, it never produces arbitrary quantum channels at each node."

"The channels achievable via the first layer of unitaries constitute only a subset of all possible channels between its input and output density matrices."

"For any unitary node in subsequent layers, there are no longer any ancillas, whereas there is at least one output qubit observed or operated on later."

"The channels achievable via each layer of unitaries then also constitute only a subset of all possible channels between its input and output density matrices."

Related Work

"The locally purified state (LPS) (Glasser and others [33]) adds to the BM some purification edges each of which partially traces over a node, and represents the most general family of quantum-inspired probabilistic models."

"The decohered Born Machine (DBM) (Miller and others [34]) adds to a subset of the virtual bonds in BM some decoherence edges that fully dephase the underlying density matrices."

"A fully-DBM, i.e., a BM all of whose virtual bonds are decohered, can be viewed as a discrete UGM (Miller and others [34])."

"Fully dephasing every virtual bond in the network gives rise to a fully DBM, which can be also viewed as a discrete UGM in the dual graphical picture."

Numerical Experiments

"To demonstrate the competing effect between dephasing and adding ancillas while accordingly increasing the bond dimension of the network, we train the unitary TTN to perform binary classification on grouped classes on three datasets of different levels of difficulty."

"On all three datasets, the performance regained after adding two ancillas across all dephasing probabilities is comparable to the performance with the no-ancilla non-dephased network."

"Due to the high computational costs with more than three ancillas added to the network, our experiments do not provide sufficient information about whether the corresponding Bayesian network in the fully dephased limit will ever reach the same level of classification performance as the non-dephased unitary TTN by increasing the number of ancillas."

"To examine to what extent the observed performance decrement may be attributed to decoherence within the network as opposed to decoherence of the data qubits, we perform the same numerical experiment on the Fashion-MNIST dataset but keep the input qubits coherent without any dephasing."

Discussion

"We investigated the competition between dephasing tensor network QML models and adding ancillas to the networks, in an effort to investigate the advantage of coherence in QML and to provide guidance in determining the number of noisy ancillas to be included in NISQ-era implementations of these models."

"We find that the performance of the two-ancilla Bayesian network, namely the fully dephased network, is comparable to that of the corresponding non-decohered unitary TTN with no ancilla, suggesting that when implementing the unitary TTN, it is favorable to add at least two arbitrarily noisy ancillas and to accordingly increase the virtual bond dimension to at least eight."

"Without any ancilla added, neither the unitary TTN nor the MERA shows a level-off performance and their performance decreases all the way until the networks are fully dephased."

Policy Gradients Using Variational Quantum Circuits [35]

This is a machine-generated summary of:

Sequeira, André; Santos, Luis Paulo; Barbosa, Luis Soares: Policy gradients using variational quantum circuits [35].

Published in: Quantum Machine Intelligence (2023).

Link to original: https://doi.org/10.1007/s42484-023-00101-8

If you want to cite the papers, please refer to the original.

For technical reasons we could not place the page where the original quote is coming from.

Abstract-Summary

"Variational quantum circuits are being used as versatile quantum machine learning models."

"We considered a variational quantum circuit composed of a low-depth hardware-efficient ansatz as the parameterized policy of a reinforcement learning agent."

"We empirically verify that such quantum models behave similarly to typical classical neural networks used in standard benchmarking environments and quantum control, using only a fraction of the parameters."

"We study the barren plateau phenomenon in quantum policy gradients using the Fisher information matrix spectrum."

Introduction

"Variational quantum circuits (VQCs) are a viable alternative since state-action pairs can be parameterized, enabling, at least in theory, a reduction in the circuit's complexity."

"This paper proposes an RL agent's policy resorting to a shallow VQC and studies its effectiveness when embedded in the Monte-Carlo-based policy gradient algorithm REINFORCE (Williams [36]) throughout standard benchmarking environments."

"In this setting, we considered a quantum RL agent that optimizes the gate fidelity in a model-free setting, learning directly from the interface with the noisy environment."

"The main contributions of this paper are: Design of a variational softmax-policy using a shallow VQC similar to or outperforming long-term cumulative reward compared to a restricted class of classical neural networks used in a set of standard benchmarking environments and the problem of quantum state preparation, using a fraction of the number of trainable parameters."

Related Work

"Most results to date focus on value-based function approximation rather than policy-based."

"Sanches and others [37] proposed a hybrid quantum-classical policy-based algorithm to solve real-world problems like vehicle routing."

"The authors suggest that the variational method could solve quantum control problems."

"Jerbi and others [38] propose a novel quantum variational policy-based algorithm achieving better performance than previous value-based methods in a set of standard benchmarking environments."

"Compared to Jerbi and others [38], this work considers a more trivial set of observables for the measurement of the quantum circuit, leading to fewer shots necessary to estimate the agent's policy and respective policy gradient."

Policy Gradients

"A known limitation of the REINFORCE algorithm is due to Monte Carlo estimates."

"Stochastically sampling the trajectories results in gradient estimators with high variance, which deteriorate the performance as the environment's complexity increases (Greensmith and others [39])."

"The REINFORCE estimator can be improved by leveraging a control variate known as baseline $b(s_t)$, without increasing the number of samples N. Baselines are subtracted from the return such that the optimization landscape becomes smooth."

Quantum Policy Gradients

"The main disadvantage is the linear dependence between the number of qubits and the number of features characterizing the agent's state and the poor representational power, at least in principle (Schuld [40])."

"A single layer is composed of two single-qubit σ_y,σ_z rotation gates per qubit, followed by a cascade of entangling gates, such that features are correlated in a highly entangled state."

"The ansatz includes 2n single-qubit rotation gates per layer, each gate parameterized by a given angle."

"If the number of features in the agent's state is larger than the number of actions, the single-qubit measurements occur only on a subset of qubits."

"With probability $1 - \delta_{()}$ and approximation error $\epsilon_{()}$, the quantum policy gradient algorithm requires a number of shots given by Similarly to Lemma 4.1, it is shown that the accuracy of the policy gradient, as a function on the total number of shots, grows logarithmically with the total number of parameters."

Performance in Simulated Environments

"The quantum agent should learn the optimal pulse sequence that maps to the state with maximum fidelity as the number of episodes increases."

"Since the environment is simulated, the qubit is prepared in the state of time step t and then fed to the variational quantum policy."

"The results show that a fully connected neural network with a single layer of 128 and 32 neurons performs reasonably better than similar architectures for the CartPole-v0 and Acrobot-v1 environments, respectively."

"One can conclude that the quantum and classical neural policies perform similarly in every environment."

"The input layer of a classical neural network is related to the number of qubits in a quantum circuit."

"One can conclude that the quantum policy has similar or even outperforming behavior compared to the classical policy with an extremely reduced total number of trainable parameters."

Quantum Enhancements

"Fisher information — The Fisher information matrix spectrum is related to the effect of barren plateaus in the optimization surface itself."

"It can be used as a measure for studying barren plateaus in maximum likelihood estimators (Karakida and others [41]), given that all the matrix entries will approach zero with the flatness of the model's landscape."

"If the model is in a barren plateau, then the eigenvalues of the matrix will approach zero (Abbas and others [5])."

"This work considers the trace and the eigenvalues' probability density of the Fisher information matrix."

"The trace will approach zero if the model is closer to a barren plateau and the eigenvalues' probability density unveils the magnitude of the associated eigenvalues."

"The Fisher information matrix of the quantum model exhibits significantly larger density in eigenvalues different from zero compared to the classical model during the entire training."

Conclusion

"The quantum-inspired policy needs fewer interactions to converge to an optimal behavior, benefiting from a reduction in the total number of trainable parameters."

"Parameter-shift rules were used to perform gradient-based optimization resorting to the same quantum model used to compute the policy."

"The Fisher information spectrum was used to study the effect of barren plateaus in quantum policy gradients."

"It would be interesting to embed the quantum Fisher information in a natural gradient optimization (Stokes and others [42]) to derive quantum natural policy gradients."

"Advanced RL models such as actor-critic or deep deterministic policy gradients (DDPG) could benefit from quantum-aware optimization."

Quantum Algorithms: Applications, Criteria and Metrics [43]

This is a machine-generated summary of:

Durán, Claudia; Carrasco, Raúl; Soto, Ismael; Galeas, Ignacio; Azócar, José; Peña, Victoria; Lara-Salazar, Sebastián; Gutierrez, Sebastián: Quantum algorithms: applications, criteria and metrics [43].

Published in: Complex & Intelligent Systems (2023).

Link to original: https://doi.org/10.1007/s40747-023-01073-9

If you want to cite the papers, please refer to the original.

For technical reasons we could not place the page where the original quote is coming from.

Abstract-Summary
"In the field of data processing and IoT communication it is possible to develop more robust solutions by combining quantum algorithms with metaheuristics."

"Said solutions can be applied in the industry and be measured using metrics associated with complexity, efficiency, processing, and accuracy."

"An extensive bibliographical review is carried out to determine which is the most efficient and effective hybrid algorithm that can be applied to a real experimental case, which aims to improve communication to reduce occupational risks."

Introduction

"If the network is vulnerable, it can lead to cyber-attacks in which communication systems are corrupted and false information can be transmitted that changes the real data [44]."

"The quantum provides a greater availability, scalability, and operativity to the balancing of the data loading in the cloud of the computing network working with IoT [45]."

"Quantum systems have advantages in the industry since they can minimize energy costs due to lower network traffic, improve performance with less delay latency, reduce the risks of failures in the information security field, and allow distributed use and analysis of IoT data in offline or limited connectivity environments [46]."

"As a solution, it is necessary to find communication methods that are more secure and capable of transmitting information in a ventilation system in which data are captured with sensors, optimized with a hybrid quantum algorithm, monitored and controlled with a more robust communication support."

Literature Review

"A search was carried out in the Web of Science, IEEE, and Scopus databases on October 20, 2022, with the keywords 'Quantum Algorithms' AND 'Metaheuristic Algorithm'."

"Results were obtained for Metaheuristic Algorithms (10,463 on the Web of Science, 5257 on IEEE, and 19,049 on Scopus), and for Quantum Algorithms (27,139 on the Web of Science, 38,736 on Scopus, and 10,483 on IEEE)."

"In third place appeared the quantum genetics and the firefly colony algorithm with 17.9%."

Materials and Methods

"Unlike classical computing, quantum computing uses superposition and entanglement, in which the quantum states of two or more objects are to be described by a single state involving all objects in the system even when the objects are spatially separated; since the electron may be in any of the infinitely many intermediate quantum states between classical states 0 and 1 [47, 48]."

"Unlike the classical genetic algorithm in which the population evolves genetically by selecting, crossing, and mutating genes; QGA uses the method of chromosome evolution based on the quantum rotating door, increasing its performance and the interference crossover that provides a greater crossover of the [49] chromosomes."

"The new quantum chromosome is obtained: wherein: The number of the population of quantum chromosomes is initialized: A measurement is performed with a group of possible solutions according to the iteration in which the execution is [50]."

Experiment

"The feasibility of improving communications in an underground mining environment with an optimized quantum genetic algorithm will be studied in the experiment."

"With the radiofrequency nodes in the tunnel, the sensor information is sent to a hub with a fixed position, and an IoT network is built."

"The information is sent from each LoRa node to the hub, which sends the data to the local server."

"Step 2: Optimization of the Quantum Genetic Algorithms."

"The "cloud" is a combination between The Things Network (TTN) server, that manages the LoRaWAN communication and computing services at a local server, wherein the services of optimization of the quantum-genetic algorithm are stored."

"The chromosome of the algorithm comes from the creation of a random vector matrix that generates the individual selection process and their genetic crossing."

Experimental Results

"GA converges rapidly in the second generation with values close to 1.5173."

"In the experiment, the following considerations were taken: A sensor package is not located at a point in the environment, at a shorter distance than another package of the same type."

"Considering the technical specifications of the sensors since the precision and accuracy can be improved for the sensor package closest to the object."

Discussion

"In a quantum algorithm, it is possible to have different tuning strategies of the quantum revolving door that could prematurely converge locally at a slow rate, in a state of stagnation [51]."

"In the optimization function, further analysis of the chromosome function of the quantum genetic algorithm is required [52]."

"It is necessary to improve the quantum genetic algorithm by analyzing other optimization functions on the chromosome, which do not use only the fitness function as the traditional genetic algorithm and which deliver solutions that have fast convergence to the local optimum."

"More research is needed on hybrid quantum algorithms, wherein metaheuristics that generate a global search and optimization methods that perform an efficient local search are combined [53]."

"Another difficulty was the creation of the database with unstructured information; it was necessary to clean and structure the data so that it could be processed with the quantum genetic optimization algorithm."

Conclusions

"Despite the fact that there are industrial applications in the literature, it has been difficult to understand how they can be applied to a real case since it is necessary to have knowledge and understanding of different fields, such as: quantum mechanics, metaheuristics, function optimization, programming, data analytics, and electrical engineering."

"Despite the multidisciplinary nature of the case study presented in this work, it was possible to integrate experimental data obtained with sensors with input parameters of the quantum genetic algorithm to obtain results that are close to the global optimum."

"New methods and protocols are required to transmit quantum information securely and in real time, so that in the future it will be possible to have a quantum internet in companies to ensure communication between the sender and receiver [54]."

"With quantum, data could be sent over long distances, with secure cryptography."

Near-Optimal Quantum Algorithms for String Problems [55]

This is a machine-generated summary of:

Akmal, Shyan; Jin, Ce: Near-Optimal Quantum Algorithms for String Problems [55].

Published in: Algorithmica (2023).

Link to original: https://doi.org/10.1007/s00453-022-01092-x

If you want to cite the papers, please refer to the original.

For technical reasons we could not place the page where the original quote is coming from.

Abstract-Summary
"We study quantum algorithms for several fundamental string problems, including Longest Common Substring, Lexicographically Minimal String Rotation, and Longest Square Substring."

"These problems have been widely studied in the stringology literature since the 1970s, and are known to be solvable by near-linear time classical algorithms."

"We give quantum algorithms for these problems with near-optimal query complexities and time complexities."

"Our algorithm is an adaptation of the algorithm by Le Gall and Seddighin (2022) for the Longest Palindromic Substring problem, but uses additional techniques to overcome the difficulty that binary search no longer applies."

Introduction

"There remain many other string problems which have near-linear time classical algorithms with no known quantum speed-up."

"We give high-level overviews of our quantum algorithms for Longest Common Substring (LCS), Minimal String Rotation, and Longest Square Substring."

"We sketch the main ideas of our improved quantum algorithm for Minimal String Rotation by comparing it to the previous best solution for this problem."

"This algorithm is inefficient in its first step, where it uses quantum minimum finding to obtain the minimum length-B prefix P. The length-B prefixes we are searching over all come from rotations of the same string s. Due to this common structure, we should be able to find their minimum more efficiently than just using the generic algorithm for minimum finding."

"Booth [56] and Shiloach [57] gave the first linear time algorithms for the Minimal String Rotation problem."

Preliminaries

"We assume the input strings can be accessed in a quantum query model [58, 59], which is standard in the literature of quantum algorithms."

"Besides the input queries, the algorithm can also apply intermediate unitary operators that are independent of the input oracles."

"The query complexity of an algorithm is the number of queries it makes to the input oracles."

"We are also interested in the time complexity of the quantum algorithms, which counts not only the queries to the input oracles, but also the elementary gates [60] for implementing the unitary operators that are independent of the input."

"Quantum random access, a classical time-T algorithm that uses random access memory can be converted into a quantum subroutine in time O(T), which can be invoked by quantum search primitives such as Grover search."

Longest Common Substring

"As noticed by Ambainis [[61], Section 6.2], such data structures have to satisfy the following requirements in order to be applicable in quantum walk algorithms."

"The reason is that, during the quantum algorithm, each operation is actually applied to a superposition of many instances of the data structure, so we would like the time complexity of an operation to have a fixed upper bound that is independent of the particular instance being operated on."

"Using this data structure to maintain the arrays in our quantum walk algorithm, we can implement the update steps and the setup steps time-efficiently."

"We dynamically maintain the solution using some data structure, which efficiently handles each update step during the quantum walk where we insert one string pair (P, Q) into (and remove one from) the current Two String Families LCP instance."

"During the quantum walk algorithm, each data structure operation is aborted after running for T time steps."

Minimal String Rotation

"Rather than work with the Minimal String Rotation problem directly, we present an algorithm for the following problem, which is more amenable to work with using our divide-and-conquer approach."

"The Minimal String Rotation problem reduces to the Maximal String Rotation problem."

"We have because the string on the left hand side is a proper prefix of the string on the right hand side."

"By assumption we have Because the string on the left hand side occurs strictly before the string on the right hand side in lexicographic order, appending any number 0s to the ends of the strings above cannot change their relative order."

"We remark that similar kinds of exclusion rules have been applied previously in parallel algorithms for Exact String Matching [62] and Minimal String Rotation [63] (under the name of "Ricochet Property" or "duel"), as well as the quantum algorithm by Wang and Ying [[64], Lemma 5.1]."

"Suppose there is an T(d)-time quantum algorithm for checking whether an O(d)-length string is a Lyndon word."

Longest Square Substring

"If we never find such a string we report that s has no square substring."

"If s has no square substring, our algorithm will never find a solution a will correctly detect that there is none."

"Because this substring is p-periodic and has length an even multiple of p, it is a square substring."

"Because it has length equal to A, it is a longest square substring as desired."

"We construct two candidate solutions, and afterwards prove that one of them is guaranteed to be a longest square substring."

"We return the largest square substring among these two."

"The longest square substring of the 0s string of length 2n is just the entire string, and has length 2n."

"Every other string in S has an odd number of 1s, and thus has longest square substring of size strictly less than 2n."

Open Problems

"We conclude by mentioning several open questions related to our work."

"A subsequent work by Childs, Kothari, Kovacs-Deak, Sundaram, and Wang [65] showed such an improvement for the decision version of Minimal String Rotation, but the question remains open for the search version."

"Ambainis' implementation [61] additionally used a skip list, and Jeffery's (non-comparison-based) implementation used a quantum radix tree [[66], Section 3.3.4]."

A Quantum-Inspired Online Spiking Neural Network for Time-Series Predictions [67]

This is a machine-generated summary of:

Yan, Fei; Liu, Wenjing; Dong, Fangyan; Hirota, Kaoru: A quantum-inspired online spiking neural network for time-series predictions [67].

Published in: Nonlinear Dynamics (2023).

Link to original: https://doi.org/10.1007/s11071-023-08655-9

If you want to cite the papers, please refer to the original.

For technical reasons we could not place the page where the original quote is coming from.

Abstract-Summary

"Spiking neural networks (SNNs) are considered the most promising new generation of artificial neural networks, due to their superior dynamic structures and low energy consumption, resembling that of the biological brain."

"The network performance of traditional spiking encoding methods is significantly affected by noise."

"This study proposes a quantum-inspired online spiking neural network (QiSNN), which combines a quantum particle swarm optimization algorithm and a Kalman filtering technique to smooth and denoise the original time-series data."

"Experimental results demonstrate that the proposed QiSNN outperforms baseline models across multiple evaluation metrics."

Introduction

"Hu and others proposed a spike-timing-dependent plasticity-based weight-quantized/binarized online learning SNN, which, when applied to the MNIST training set, reduced storage requirements while improving computational efficiency, demonstrating the advantages of SNNs for online learning [68]."

"Shi and others proposed concentrating quantum-inspired neurons and combined quantum data representations and convolutional operations to construct quantum convolutional neural networks, which can process high-dimensional data and surpass classical models in more practical learning tasks [69]."

"A quantum-inspired online spiking neural network (QiSNN) is proposed in this study to limit the detrimental effects of noise on the SNN model."

"Our contributions include the following: The proposed QiSNN combines quantum particle swarm optimization (PSO) with a Kalman filtering algorithm to reduce the detrimental effects of noise information in the original spiking neural network data; A threshold selection technique is developed to simplify and optimize the process of identifying similarities between neurons; The proposed QiSNN model is applied to air quality estimation, demonstrating its advantages for short-term predictions."

Preliminary Work

"The Kalman filter, proposed by R.E. Kalman in the 1960s, utilizes the state equation of a linear system to optimally estimate a system state based on input and output data [70]."

"The optimal state estimate can be obtained from the state update equation: where K is the Kalman gain, which characterizes the weight of model prediction errors versus measurement errors in the state-optimal estimation process."

"OeSNNs are distinguished from eSNNs by two key components: the input and output layer structure, and an ability to classify data stream values by adjusting the number of neurons."

"After the initial data are supplied, the input data can be encoded in the encoding layer using a variety of techniques, the most common being Gaussian receptive field (GRF) population encoding."

"The output layer then categorizes these data as a repository, and input neurons are joined to output neurons using synapses."

Quantum-Inspired Spiking Neural Network

"The data processing layer in the QiSNN uses a quantum PSO algorithm and a Kalman filter to smooth and denoise the original input time series, reduce the effects of noise on the initial input data, and improve time-series prediction outcomes."

"The time-series data intersect this Gaussian function in the GRF, and the value of this intersection affects the time at which the input neuron transmits spikes (or the order in which all input neurons transmit spikes)."

"Due to the complexity and diversity of time-series prediction applications, it is difficult to identify a set threshold to determine the similarity of neurons for the effective classification of the original data."

"The direct selection of a threshold influences similarity evaluation, and the process has often been insufficiently described."

"The proposed method produces a more practical and efficient dynamic threshold, which can (to some extent) reflect the degree of dispersion in the original data and adapt to different backgrounds and problems."

Qisnn-Based Air Quality Prediction

"The measured one-, three-, and six-hour accuracy rates are respectively 88.9%, 79.9%, and 67.7% for QiSNN, and 77.4%, 76.7%, and 71.0% for CEeSNN predictions at the monitoring site in London Bloomsbury."

"QiSNN produces one-, three-, and six-hour predictions for London Marylebone data with accuracies of 91.0%, 82.2%, and 73.1%, respectively."

"QiSNN performs best overall among the one-, three-, and six-hour predictions made at the London North Kensington monitoring site, with accuracy rates of 90.1%, 80.3%, and 70.6%, respectively."

"Accuracy rates for QiSNN applied to one-, three-, and six-hour London Marylebone predictions are 88.4%, 83.2%, and 76.0%, respectively."

"QiSNN produces one-, three-, and six-hour accuracies of 97.1%, 95.3%, and 88.8%, respectively, for the London North Kensington data, among which the accuracies of one- and three-hour predictions are the highest among the five models."

Conclusion

"The quantum PSO algorithm utilizes quantum advantages to enable all particles in the population to appear in the search space with a certain probability."

"We proposed a quantum-inspired online spiking neural network (QiSNN), which combines and takes full advantage of quantum PSO and Kalman filtering algorithms, enhancing network performance by reducing the effects of noise in the original data."

"Comparing the proposed model with four baseline algorithms, as applied to air quality predictions, revealed that QiSNN provides clear advantages, particularly over short time periods."

"The proposed QiSNN consists of data processing, input, and output layers."

"The potential of QiSNN for time-series predictions [71] will be explored further (beyond air quality forecasting), including the prospect of multimodal emotion prediction."

Quantum Neural Network Autoencoder and Classifier Applied to an Industrial Case Study [72]

This is a machine-generated summary of:

Mangini, Stefano; Marruzzo, Alessia; Piantanida, Marco; Gerace, Dario; Bajoni, Daniele; Macchiavello, Chiara: Quantum neural network autoencoder and classifier applied to an industrial case study [72].

Published in: Quantum Machine Intelligence (2022).

Link to original: https://doi.org/10.1007/s42484-022-00070-4

If you want to cite the papers, please refer to the original.

For technical reasons we could not place the page where the original quote is coming from.

Abstract-Summary

"In these early practical uses of quantum computers, it is relevant to develop algorithms that are useful for actual industrial processes."

"This work represents one of the first attempts to integrate quantum computing procedures in a real-case scenario of an industrial pipeline, in particular using actual data coming from physical machines, rather than pedagogical data from benchmark datasets."

Introduction

"To overcome the problems due to the limited number of qubits available and to the absence of efficient error correction techniques, several proof-of-principle demonstrations have been carried out by focusing on so-called variational quantum algorithms, characterized by a hybrid approach in which the quantum processing units (QPU) is seen as an accelerator alongside the classical CPU (McClean et al. [73]; Cerezo et al. [74]; Bharti et al. [75])."

"Most quantum machine learning algorithms are based on parametrized quantum circuits, and leverage an approach in which the optimization over the variational parameters is done on the classical CPU (Benedetti et al. [76]; Mangini et al. [77])."

"We propose the application of a newly formulated quantum pipeline comprising a quantum autoencoder algorithm (Romero et al. [78]; Bravo-Prieto [79]; Lamata et al. [80]; Khoshaman et al. [81]) followed by a quantum classifier, applied to real data coming from a first stage water/oil separator of one of Eni's oil treatment plant."

Case Study

"Due to the limitation in the complexity of the problems that can currently be faced with quantum computing, we will focus on a simplified problem, involving only four variables, which are: the oil level (LIC), the oil output flow (FT), the pressure (PI), the opening of the oil output valve (FRC)."

"We run a binary clustering algorithm on the initial variables, in order to identify two categorical states, named as "Class A" and "Class B", and then used these categorical states as the labels for the classification task."

"The latent vector from the encoder is used as input for the classifier, which is trained to correctly predict the "Class A" and "Class B" states."

"This algorithm takes as input the desired number of clusters, in our case two, and tries to split the data in groups of equal variance."

Neural Network Autoencoder

"Using this dataset, called training set, a neural network can be trained in a supervised fashion to learn the relationship between the input variables and the expected classification results."

"This leads to a situation where too many input variables are available in the dataset, and it is often ineffective to directly feed them into the neural network classifier."

"An autoencoder is a neural network composed of two modules, called encoder and decoder, designed in such a way that the subsequent application of the encoder and the decoder to the input data results into an output that is as close as possible to the input, i.e., the discrepancy between output and input is minimized."

"The neural network is trained in an unsupervised fashion in order to generate an output that is as close as possible to the input."

Quantum Data Compression

"In order to use a quantum pipeline to analyze the classical data coming from the sensors, we need to encode such data on a quantum state to be used as the input of the quantum autoencoder."

"While it is known from the recent literature (Abbas et al. [5]; Lloyd et al. [82]; Schuld et al. [83]; Gil Vidal and Theis [84]; LaRose and Coyle [85]; Mitarai et al. [86]) that choosing a good encoding scheme is of key importance to ensure good expressivity and representation power of variational quantum algorithms, there is still no standard procedure to do so."

"The goal of the encoder is to disentangle the two systems in such a way that one of them, i.e., the trash system, goes to the fixed reference state, and the other contains all the original information of from the full quantum state."

"Acting with the decoder on the compressed state yields the original state Thus, suppose having compressed the information stored in the quantum state of a composite system into one of its subsystems."

Experiments and Results

"We discuss the experiments implementing the classical and quantum data analysis approaches described above for the data compression and classification tasks."

"We expect that, if the compressed vector is a suitable representation of the input data, a classification algorithm would be able to achieve very good performances."

"Single qubit quantum classifier Once the quantum autoencoder has been trained to learn a compressed representation of the original information, the compressed quantum state can be used as input for a classification task."

"A quantum classifier is made of two parts: a trainable parametrized operation, U, which tries to map inputs belonging to different classes in two distant regions of the Hilbert space; and a final measurement, which is used to extract and assign the label."

"The quantum classifier, given its relatively simple structure, learns essentially a straight cut of the data in this region, thus committing some labeling errors."

Summary and Outlook

"We have presented a direct comparison between quantum and classical implementations of a neural network autoencoder, followed by a classifier algorithm, applied to sample real data coming from one of Eni's plants, in particular from a first stage separator."

"We provided, on the other hand, a successful proof-of-concept demonstration that an original quantum autoencoder and a quantum classifier can actually reach the same level of accuracy as standard classical algorithms, on a data set that is sufficiently low dimensional to be handled on actual near-term quantum devices."

"It is worth emphasizing that the quantum autoencoder allows to obtain results that are quantitatively comparable to the classical algorithm by using only six parameters instead of 16, thus displaying an increased efficiency in terms of number of trainable parameters already reached on NISQ devices."

"We believe these results take the first foundational steps towards the application of usable quantum algorithms on NISQ devices for industrial data."

Quantum Encryption and Generalized Shannon Impossibility [87]

This is a machine-generated summary of:

Lai, Ching-Yi; Chung, Kai-Min: Quantum encryption and generalized Shannon impossibility [87].

Published in: Designs, Codes and Cryptography (2019).
Link to original: https://doi.org/10.1007/s10623-018-00597-3

If you want to cite the papers, please refer to the original.
For technical reasons we could not place the page where the original quote is coming from.

Abstract-Summary
"We also give a systematic study of information-theoretically secure quantum encryption with two secrecy definitions."

"We show that the weaker one implies the stronger but with a security loss in d, where d is the dimension of the encrypted quantum system."

Introduction

"Encryption schemes are typically considered as a computational primitive, since an information-theoretically secure (ITS) symmetric key encryption scheme can only securely encrypt messages of length at most the length of the secret key by Shannon's impossibility result [88]."

"We start by revisiting the quantum Shannon impossibility for quantum encryption using classical keys."

"There is a lower bound on the entropy of keys that if an encryption scheme that perfectly encrypts n qubits using a classical key drawn from K, then where H is the Shannon entropy."

"As an application, we can prove the security of an encryption with respect to this weak notion when the secrecy loss is not that important."

"We will also prove a Shannon impossibility result for this weak notion of security."

Preliminaries

"A quantum system will bc denoted by a capital letter and its corresponding complex Hilbert space will be denoted by the corresponding calligraphic letter."

"N-fold Pauli matrices form a basis for linear operators on n qubits."

"We will use a subscript to specify which subsystem a vector belongs to or an operator operates on."

"When A is classical, is called a classical-quantum (cq) state."

Information-Theoretically Secure Quantum Encryption

"A symmetric-key quantum encryption scheme is defined as follows."

"Following that, an information-theoretic security notion of entropy security was defined (in a classical setting) by Russel and Wang [89]."

"Dodis and Smith [90] discussed the security notions more generally and showed that entropy security is equivalent to a security notion of indistinguishability."

"Entropic security and entropic indistinguishability are also equivalent in the quantum settings [91, 92]."

"We show that an weak ITS encryption scheme is also ITS but with an additional security loss in the dimension of the message system to be encrypted."

Discussion

"We studied quantum encryption with imperfect correctness and imperfect secrecy."

"The quantum Shannon impossibility results were generalized to the case with imperfect secrecy and imperfect correctness for both notions of security."

"Our Shannon impossibility results for weak ITS scheme also agrees that roughly 2n bits are necessary to encrypt n qubits."

A Novel Quantum-Inspired Solution for High-Performance Energy-Efficient Data Acquisition from IoT Networks [93]

This is a machine-generated summary of:

Bhatia, Munish; Sood, Sandeep; Sood, Vaishali: A novel quantum-inspired solution for high-performance energy-efficient data acquisition from IoT networks [93].

Published in: Journal of Ambient Intelligence and Humanized Computing (2020).

Link to original: https://doi.org/10.1007/s12652-020-02494-x

If you want to cite the papers, please refer to the original.

For technical reasons we could not place the page where the original quote is coming from.

Abstract-Summary

"Conspicuously, this study proposes a Quantum Computing inspired technique of temporal space optimization for real-time big IoT applications."

"Quantification of IoT sensors is performed in terms of Sensors of Interest (SoI) and Degree of Aptness (DoA) measure to minimize IoT sensor-space in real-time."

"2 key performance indicators in terms of Data Similarity Analysis and Energy Efficiency are estimated for optimized efficacy."

"To evaluate the presented technique, numerous simulations are performed in real-time scenario of vehicular traffic determination over 1 km of Regional National Highway using 70 WiSense nodes comprising of noise sensors, vibration sensors, and Raspberry Pi device."

"Results registered depict that significant improvements are registered for the presented technique in terms of temporal effectiveness, and performance parameters like Accuracy, Correlation Analysis, and Reliability."

Introduction

"With the enhanced capabilities to insert intelligence everywhere, IoT devices are being used to map the physical world where field-specific operations are happening and the software world where data processing and decision-making is performed (Liu et al. [94])."

"With such a large IoT environment, realizing a common goal of optimal data accuracy in ubiquitous time-sensitive applications is a challenging perspective."

"The minimization of data in an IoT environment or IoT-space in a given time frame for optimal perception in another issue of concern for generating effective results in real-time."

"Inspired from the beneficial aspects of QCIO, the work presented in this paper is focused on quantamizing the temporal aspects of IoT technology in terms of sensor-specific parameter determination, defining IoT temporal space, sensor selection in real-time, and generating time-sensitive results."

"When time sensitiveness is concerned, researchers are still developing solutions for optimal IoT architectures to provide services with minimal delay and maximum performance."

Motivational Works

"Authors have proposed a semantic reasoning-based system for performing operations in real-time scenarios of IoT. The proposed framework was evaluated using different semantic data models in which promising results were achieved."

"Boveiri et al. [95] presented an effective approach for solving the task-scheduling problem in a cluster computing environment."

"The experimental simulation depicted the enhanced performance of the proposed algorithm in comparison to other scheduling techniques."

"Boveiri [96] presented an enhanced version of the cuckoo optimization algorithm (COA) for solving the static task scheduling problem in a cluster computing environment."

"Boveiri et al. [97] proposed a high-performance approach based on the Max-Min Ant System, an enhanced version of ant colony optimization for tackling the task scheduling problem in a multiprocessor environment."

"Zhao et al. [98] proposed a geometric-constrained multi-view image matching approach for computing cost value of energy function utilizing the semi-global optimization method."

Proposed Mechanism

"Optimization of IoT temporal space comprises of identification of the optimal IoT sensors for the accurate perception of real-time data streams."

"With consideration to these aspects, this research defines a minimization problem of optimizing IoT temporal space for sensing real-time data values."

"To define the minimized IoT temporal space, every sensor is assessed based on the above-mentioned parameters."

"This section presents the dynamic QCIO algorithm for optimizing minimal-IoT Temporal Space for acquiring real-time data from multiple IoT sensors in the application environment."

"QCIO is based on quantum computing formalization to minimize IoT sensor space for the elimination of false data acquisition based on sensor-specific parameters."

"Minimization of IoT temporal space not only enables accurate data acquisition in real time but also provide effective generation of application-oriented data sets."

Experimental Implementation

"It can be seen that for noise-related data, the proposed QCIO technique was able to acquire accurate data segments in an average time of 15.69 s. In comparison to this, ACO registered accurate data segments in 26.69 s and PSO in 23.47 s. GBiO was able to acquire datasets in a maximal time of 20.36 s. On the other hand, the presented technique of QCIO was able to outperform other models for vibration datasets."

"ACO was able to compute accurate data segments in an average of 35.69 s, PSO in 29.15 s and GBiO in 27.84 s. Based on the results registered, it is inferred that in the recent scenario, the stated technique of QCIO is temporally efficient for the reduced IoT temporal space optimization."

Comparative Analysis

"The comparative study of the discussed QCIO model with other existing models is elaborated in this section."

"5 related works of literature are mentioned for the comparative assessment."

"These are Zhao et al. [99], Liu et al. [100], Dey et al. [101], Dey et al. [101], and Bhatia et al. [46]."

"Purposes, 8 different parameters are utilized."

Discussion and Conclusions

"This paper presented a Quantum Computing Inspired Optimization (QCIO) approach for minimizing the IoT-Temporal space."

"In this research (1) A novel QCIO algorithm was proposed for accurate data acquisition by IoT devices in real-time. (2) QCIO-technique was presented to represent the overall temporal space in terms of Sensors of Interest (SoI) and Degree of Aptness (DoA) measure for IoT devices. (3) The multi-object data sensation model was proposed for accuracy enhancement of data acquisition."

"For validation purposes, numerous simulations were performed for data acquisition in a vehicular environment using WiSense nodes, Raspberry Pi, and quantum simulators."

"Based on results, it was concluded that the presented method has been able to enhance data accuracy in a multi-sensor IoT environment with minimal delay."

Quantum-Classical Reinforcement Learning for Decoding Noisy Classical Parity Information [102]

This is a machine-generated summary of:

Park, Daniel K.; Park, Jonghun; Rhee, June-Koo Kevin: Quantum-classical reinforcement learning for decoding noisy classical parity information [102].

Published in: Quantum Machine Intelligence (2020).

Link to original: https://doi.org/10.1007/s42484-020-00019-5

If you want to cite the papers, please refer to the original.

For technical reasons we could not place the page where the original quote is coming from.

Abstract-Summary

"Learning a hidden parity function from noisy data, known as learning parity with noise (LPN), is an example of intelligent behavior that aims to generalize a concept based on noisy examples."

"We show that a naive application of the quantum LPN algorithm to classical data encoded in an equal superposition state requires an exponential sample complexity."

"We then propose a quantum-classical reinforcement learning algorithm to solve the LPN problem for classical data and demonstrate a significant reduction in the sample complexity compared with the naive approach."

"Simulations with a hidden bit string of length up to 12 show that the quantum-classical reinforcement learning performs better than known classical algorithms when the sample complexity, run time, and robustness to classical noise are collectively considered."

Introduction

"The ability to exhibit the advantage of quantum learning, especially when training examples are classical, remains an interesting and important open problem."

"We show that a naive application of the quantum LPN algorithm to classical data requires an exponential amount of examples (i.e., training samples) or computing resources, thereby nullifying the quantum advantage."

"We then propose a quantum-classical hybrid algorithm based on the reinforcement learning framework for solving the LPN problem in the absence of the quantum oracle."

"The proposed algorithm uses noisy classical samples to prepare an input quantum state that is compatible with the original quantum LPN algorithm."

"Based on the outcome of the quantum algorithm, a reward is classically evaluated and an action is chosen by a greedy algorithm to update the quantum state in the next learning cycle."

Learning Parity with Noise

"In the quantum LPN algorithm introduced in Cross and others [103], a quantum oracle implements a unitary transformation on the computational basis states and returns the equal superposition of $|x\rangle|f(x)\rangle$ for all possible inputs x. By applying Hadamard gates to all qubits at the query output, the learner acquires an entangled state: Thus, whenever the label (last) qubit is 1 (occurs with probability 1/2), measuring data (first n) qubits in their computational bases reveals s. Note that this algorithm is very similar to the Bernstein-Vazirani (BV) algorithm (Bernstein and Vazirani [104]), except that in the BV problem the learner can choose an example in each query and the input state of the label qubit is prepared in $|-\rangle$."

"The quantum LPN algorithm requires the extra post-selection step since the input of the label qubit is prepared in $|0\rangle$."

"Learning the hidden parity function from noiseless examples is efficient for both classical and quantum oracles."

"The advantage of having a quantum oracle for solving an LPN problem was demonstrated experimentally with superconducting qubits in Ristè and others [105]."

Naive Application of Quantum Algorithm to Classical Data

"The measurement outcome of 1, which occurs with the probability of 1/2, is post-selected to leave the query register qubits in: From the above state, the probability to guess s correctly is: which is exponentially small in n. However, this result also implies that even when the quantum oracle outputs only a fraction of all possible examples as an equal superposition state, and the noise does not act coherently on all x_j as in Cross and others [103] and Ristè and others [105], the LPN problem can still be solved."

"The quantum state can be prepared by guessing the quantum oracle of the quantum LPN algorithm, and inserting the output state of the oracle as an input to quantum random access memory (QRAM) (Giovannetti and others [106, 107]; Hong and others [108]; Arunachalam and others [109]; Park and others [110]) to update its entries according to real data."

"The circuit-based QRAM introduced in Ref. (Park and others [110]) can use flip-register-flop processes to update an output of a guessed quantum oracle with real data using the number of steps that increases at least linearly with the number of samples."

Reinforcement Learning

"The state in each iteration is the guessed bit string after performing the usual quantum LPN algorithm."

"The greedy algorithm then selects the guessed bit string that maximizes the reward, and uses it to construct the guessed quantum oracle of the next iteration."

"In the R-LPN algorithm, the time complexity is again dominated by state preparation, for which the number of steps increases at least linearly with the number of samples as mentioned in the previous section."

"We first simulate an R-LPN algorithm with a slight modification, which is intended to save the memory and time cost for storing all M sets of guessed parity bits to calculate their Hamming distances with respect to the real parity bits."

"We compare the performance of this modified algorithm to the originally proposed R-LPN by analyzing the success probability with respect to the number of samples via simulations."

Robustness to Pauli Errors

"Since the R-LPN algorithm performs the measurement in the σ_z basis, it is not affected by any error that effectively appears as unwanted phase rotations at the end of the quantum circuit."

"When the R-LPN algorithm is completed, it outputs the same final state as in Cross and others [103] with high probability."

"The above result can be directly applied to our algorithm for bit-flip errors on the final state."

Conclusion

"We developed a quantum-classical hybrid algorithm for solving the LPN problem with classical examples."

"The reinforcement learning significantly reduces both the sample and the time cost of the quantum LPN algorithm in the absence of the quantum oracle."

"Whether the known advantage of oracle-based quantum algorithms can be retained in the absence of the quantum oracle is an interesting open problem."

"We showed that for the LPN problem, quantum advantage can be achieved with the integration of classical reinforcement learning."

"Our results motivate future works to employ similar strategies to known oracle-based quantum algorithms in order to extend their applicability to classical data."

"Extending the idea of the quantum-classical reinforcement learning to the learning with errors problem (Grilo and others [111]) would be an interesting future work."

Network Attack Detection Scheme Based on Variational Quantum Neural Network [112]

This is a machine-generated summary of:

Gong, Changqing; Guan, Weiqi; Gani, Abdullah; Qi, Han: Network attack detection scheme based on variational quantum neural network [112].

Published in: The Journal of Supercomputing (2022).

Link to original: https://doi.org/10.1007/s11227-022-04542-z

If you want to cite the papers, please refer to the original.

For technical reasons we could not place the page where the original quote is coming from.

Abstract-Summary

"With the rapid development of quantum machine learning, variational quantum neural network (VQNN) has demonstrated quantum advantages in classification problems."

"The intrusion detection system (IDS) based on quantum machine learning has higher accuracy and efficiency than the IDS based on traditional machine learning."

"We propose a intrusion detection scheme based on VQNN, which is composed of variational quantum circuit (VQC) and classical machine learning (ML) strategy."

"In order to verify the effectiveness of the scheme, we used the VQNN model and some classic ML models (Such as artificial neural network, support vector machines, K-Nearest Neighbors, Naive Bayes, decision tree) to conduct comparative experiments."

Introduction

"Quantum neural network (QNN) is a new research field which combines the traditional ANN models with quantum information and quantum computing concepts."

"For certain specific problems, such as sampling from difficult-to-simulate probability distributions and solving the problem of vanishing gradients in quantum neural networks, NISQ equipment [113] is expected to provide higher computational advantages than classic supercomputers."

"Liang [114] proposed a hybrid quantum-classical neural network (Res-HQCNN) with deep residual learning."

"The hybrid neural network of quantum circuit based on classical convolutional neural network (CNN) is used to solve the problem of image classification."

"In order to use quantum effect and quantum advantage to solve the problem of intrusion detection, in this work, we propose a network intrusion detection scheme based on VQNN, a hybrid quantum classical computing model composed of VQC that can be executed on current NISQ equipment combined with machine learning strategies, and through iterate to complete parameter optimization."

Preliminaries

"The so-called quantum superposition state is uncertain when it is not measured."

"In quantum computing, especially in the calculation model of quantum circuits, a quantum gate (or quantum logic gate) is a basic quantum circuit that operates qubits."

"It is the basis of quantum circuits, just like the relationship between traditional logic gates and general digital circuits [115]."

"An important application of H gate is the initialization of quantum states, namely preparation of the uniform superposition of states."

Variational Quantum Neural Network

"The process of quantum neural network is similar to classical neural network, including data preprocessing, quantum state encoding, training network, optimizing parameters and obtaining the best probability of results."

"The network uses qubit registers to parameterize the input of quantum circuit and obtains the network output distribution of bit strings which occur with frequency (probability) through measurement operation, i.e., to establish the completely positive map between the two through quantum neural network."

"After the classical data are successfully encoded into quantum state, we need to consider using the existing qubits or adding some auxiliary qubits to design the quantum circuits."

"It is a kind of tunable quantum circuit which optimizes parameters through the iterative process of classical computer, we can regard these adjustable parameters as weights of artificial neural network."

Experiment and Analysis

"The experimental settings are as follows, the experimental platform is TensorFlow and Google's open source quantum machine learning platform TFQ; the system is Windows 10 professional edition; the python version is Python-3.6; The processor is Intel (R) Core (TM) i7-6700HQ CPU (8 core) @ 2.60 Hz 2.59 GHz and memory is 16 GB; the quantum circuit is a 5-qubits quantum circuit, corresponding to the number of characteristic attributes of the classical data to be encoded and the quantum circuit depth is 8; during training, we use the Adam optimizer with a learning rate of 0.01."

"In order to further verify the overall performance of the VQNN model, we conducted further experiments, using the VQC parameters trained by TFQ as the quantum gate parameters of the real quantum circuit and randomly selecting one hundred experimental data for testing."

Conclusion and Future Work

"We have shown that the proposed VQNN model can be used for network intrusion detection, i.e., can be trained the proposed VQNN to obtain the optimal variable quantum parameter of detection model, then use the detection model to perform network attack detection."

"This work provides a new idea to associate quantum computing with classical machine learning."

"Unlike the amplitude coding scheme, variational coding does not take full advantage of quantum computing, although we can verify the feasibility of applying the quantum circuit of variational encoding for network intrusion detection."

"In order to obtain an ideal quantum circuit with much fewer parameters, we can try to reduce the complexity of the parameters by using amplitude encoding."

"Future work includes applying amplitude encoding scheme to more complex data sets and variational quantum circuits to solve more challenging problems."

Quantum Convolutional Neural Network for Classical Data Classification [116]

This is a machine-generated summary of:

Hur, Tak; Kim, Leeseok; Park, Daniel K.: Quantum convolutional neural network for classical data classification [116].

Published in: Quantum Machine Intelligence (2022).

Link to original: https://doi.org/10.1007/s42484-021-00061-x

If you want to cite the papers, please refer to the original.

For technical reasons we could not place the page where the original quote is coming from.

Abstract-Summary

"We benchmark fully parameterized quantum convolutional neural networks (QCNNs) for classical data classification."

"We propose a quantum neural network model inspired by CNN that only uses two-qubit interactions throughout the entire algorithm."

"We investigate the performance of various QCNN models differentiated by structures of parameterized quantum circuits, quantum data encoding methods, classical data pre-processing methods, cost functions and optimizers on MNIST and Fashion MNIST datasets."

"Since the QCNN algorithm presented in this work utilizes fully parameterized and shallow-depth quantum circuits, it is suitable for Noisy Intermediate-Scale Quantum (NISQ) devices."

Introduction

"Motivated by the benefits of CNN and the potential power of QML, quantum convolutional neural network (QCNN) algorithms have been developed (Cong and others [23]; Kerenidis and others [117]; Liu and others [118]; Henderson and others [119]; Chen and others [120]; Li and others [121]; MacCormack and others [122]; Wei and others [123]; Mangini and others [77])."

"Previous constructions of QCNN have reported success in developing efficient quantum arithmetic operations that exactly implement the basic functionalities of classical CNN or in developing parameterized quantum circuits inspired by key characteristics of CNN."

"The study of fully parameterized QCNN for performing pattern recognition, such as classification, on classical data is missing."

"We present a fully parameterized quantum circuit model for QCNN that solves supervised classification problems on classical data."

"We benchmark the performance of the parameterized QCNN with respect to several variables, such as quantum data encoding methods, structures of parameterized quantum circuits, cost functions, and optimizers using two standard datasets, namely MNIST and Fashion MNIST, on Pennylane (Bergholm and others [124])."

Theoretical Framework

"With amplitude encoding, a quantum computer can represent exponentially many classical data."

"The computational overhead of amplitude encoding motivates qubit encoding, which uses a constant quantum circuit depth while using O(N) number of qubits."

"In previous sections, the amplitude encoding is advantageous when the quantum circuit width (i.e., the number of qubits) is considered while the qubit encoding is advantageous when the quantum circuit depth is considered."

"These two encoding schemes represent the extreme ends of the quantum circuit complexities for loading classical data into a quantum system."

"Let us denote the number of qubits in each independent block that amplitude-encodes classical data by m. Then, each block can encode $O(2^m)$ classical data."

"The hybrid encoding algorithms use fewer number of qubits than the qubit encoding and use shallower quantum circuit depth than the amplitude encoding."

Benchmark Variables

"We introduce a set of convolutional and pooling ansatz (i.e., parameterized quantum circuit templates) used in our QCNN models."

"Parameterized quantum circuits for convolutional layers in QCNN are composed of different configurations of single-qubit and two-qubit gate operations."

"Circuits 2, 3, 4, 5, 7, and 8 are taken from the work by Sim and others [125] which includes the analysis on expressibility and entangling capability of four-qubit parameterized quantum circuits."

"We modified these quantum circuits to two-qubit forms to utilize them as building blocks of the convolutional layer, which always consists of two qubits."

"Similar to the choice of ansatz for the convolutional filter, there exists a variety of choices of two-qubit circuits of the pooling layer."

"For the convolutional circuit 9, we test two QCNN constructions, with and without the parameterized two-qubit circuit in the pooling layer."

Simulation

"The simulation results demonstrate that for the case of two ansatze tested the classification performance between open- and periodic-boundary QCNN circuits are similar."

"The CNN results are only comparable to that of the qubit and dense encoding cases which requires 8 and 16 classical input nodes, respectively."

"For hybrid and amplitude encoding, which require relatively simpler data pre-processing, the number of nodes in the CNN input layer is too large to be trained with a small number of parameters as in QCNN."

"This implies that the QCNN models not only achieve higher classification accuracy than the CNN models under similar training conditions but also are less sensitive to the random initialization of the free parameters."

"We show such data for two cases in MNIST data classification: circuit 9b and qubit encoding with autoencoding, and circuit 9b and dense encoding with PCA."

Conclusion

"The QCNN algorithm can be tailored with many variables such as the structure of parameterized quantum circuits (i.e., ansatz) for convolutional filters and pooling operators, quantum data encoding methods, classical data pre-processing methods, cost functions and optimizers."

"To improve the utility of QCNN for classical data, we also introduced new data encoding schemes, namely hybrid direct encoding and hybrid angle encoding, with which the exchange between quantum circuit depth and width for state preparation can be configured."

"The QCNN models tested in this work operated with a small number of free parameters, ranging from 12 to 51."

"The comparison between QCNN and CNN is only valid for qubit and dense encoding cases in which the number of input qubits grows linearly with the dimension of the input data."

"Testing the classification performance as the QCNN models grow bigger remains an interesting future work."

References

1. Thanasilp S, Wang S, Nghiem NA, Coles P, Cerezo M (2023) Subtleties in the trainability of quantum machine learning models. Quantum Mach Intell. https://doi.org/10.1007/s42484-023-00103-6
2. Pesah A, Cerezo M, Wang S, Volkoff T, Sornborger AT, Coles PJ (2021) Absence of barren plateaus in quantum convolutional neural networks. Phys Rev X 11(4):041011
3. Sharma K, Cerezo M, Cincio L, Coles PJ (2022) Trainability of dissipative perceptron-based quantum neural networks. Phys Rev Lett 128(18):180505
4. Liu Z, Yu L-W, Duan L-M, Deng D-L (2021) The presence and absence of barren plateaus in tensor-network based machine learning. arXiv. https://arxiv.org/abs/2108.08312
5. Abbas A, Sutter D, Zoufal C, Lucchi A, Figalli A, Woerner S (2021) The power of quantum neural networks. Nat Comput Sci 1(6):403–409
6. Haug T, Kim MS (2021) Optimal training of variational quantum algorithms without barren plateaus. arXiv. https://arxiv.org/abs/2104.14543

7. Kieferova M, Carlos OM, Wiebe N (2021) Quantum generative training using r\'enyi divergences. arXiv. https://arxiv.org/abs/2106.09567
8. Kiani BT, De Palma G, Marvian M, Liu Z-W, Lloyd S (2021) Quantum earth mover's distance: a new approach to learning quantum data. arXiv. https://arxiv.org/abs/2101.03037
9. Tangpanitanon J, Thanasilp S, Dangniam N, Lemonde M-A, Angelakis DG (2020) Expressibility and trainability of parametrized analog quantum systems for machine learning applications. Phys Rev Res 2(4):043364
10. Thanasilp S et al (2022) Exponential concentration and untrainability in quantum kernel methods. arxiv. https://arxiv.org/abs/2208.11060
11. Cerezo M, Arrasmith A, Babbush R, Benjamin SC, Endo S, Fujii K, McClean JR, Mitarai K, Yuan X, Cincio L, Coles PJ (2021) Variational quantum algorithms. Nat Rev Phys 3:625–644. https://doi.org/10.1038/s42254-021-00348-9
12. McClean JR, Boixo S, Smelyanskiy VN, Babbush R, Neven H (2018) Barren plateaus in quantum neural network training landscapes. Nat Commun 9(1):1–6
13. Cerezo M, Sone A, Volkoff T, Cincio L, Coles PJ (2021) Cost function dependent barren plateaus in shallow parametrized quantum circuits. Nat Commun 12:21728. https://doi.org/10.1038/s41467-021-21728-w
14. Larocca M, Czarnik P, Sharma K, Muraleedharan G, Coles PJ, Cerezo M (2021) Diagnosing barren plateaus with tools from quantum optimal control. arXiv. https://arxiv.org/abs/2105.14377
15. Marrero CO, Kieferová M, Wiebe N (2020) Entanglement induced barren plateaus. PRX Quant 2(4):040316
16. Patti TL, Najafi K, Gao X, Yelin SF (2021) Entanglement devised barren plateau mitigation. Phys Rev Res 3(3):033090
17. Holmes Z, Sharma K, Cerezo M, Coles PJ (2021) Connecting ansatz expressibility to gradient magnitudes and barren plateaus. PRX Quant 3:010313
18. Holmes Z, Arrasmith A, Yan B, Coles PJ, Albrecht A, Sornborger AT (2021) Barren plateaus preclude learning scramblers. Phys Rev Lett 126(19):190501
19. Huembeli P, Dauphin A (2021) Characterizing the loss landscape of variational quantum circuits. Quantum Sci Technol 6:025011
20. Zhao C, Gao X-S (2021) Analyzing the barren plateau phenomenon in training quantum neural network with the zx-calculus. arXiv. https://arxiv.org/abs/2102.01828
21. Wang S, Fontana E, Cerezo M, Sharma K, Sone A, Cincio L, Coles PJ (2021) Noise-induced barren plateaus in variational quantum algorithms. Nat Commun 12(1):1–11
22. Farhi E, Neven H (2018) Classification with quantum neural networks on near term processors. arXiv. https://arxiv.org/abs/1802.06002
23. Cong I, Choi S, Lukin MD (2019) Quantum convolutional neural networks. Nat Phys 15(12):1273–1278. https://www.nature.com/articles/s41567-019-0648-8
24. Beer K, Bondarenko D, Farrelly T, Osborne TJ, Salzmann R, Scheiermann D, Wolf R (2020) Training deep quantum neural networks. Nat Commun 11:1. https://doi.org/10.1038/s41467-020-14454-2
25. Bausch J (2020) Recurrent quantum neural networks. arXiv. https://arxiv.org/abs/2006.14619
26. Havlíček V, Córcoles AD, Temme K, Harrow AW, Kandala A, Chow JM, Gambetta JM (2019) Supervised learning with quantum-enhanced feature spaces. Nature 567(7747):209–212
27. Liao H, Convy I, Yang Z, Whaley K (2023) Birgitta Decohering tensor network quantum machine learning models. Quantum Mach Intell. https://doi.org/10.1007/s42484-022-00095-9
28. Shi Y, Duan L, Vidal G (2006) Classical simulation of quantum many-body systems with a tree tensor network. Phys Rev A 74:022320. https://doi.org/10.1103/PhysRevA.74.022320
29. Vidal G (2007) Entanglement renormalization. Phys Rev Lett 99:1–4. https://doi.org/10.1103/PhysRevLett.99.220405
30. Stoudenmire EM (2018) Learning relevant features of data with multi-scale tensor networks. Quantum Sci Technol 3:034003. https://doi.org/10.1088/2058-9565/aaba1a

31. Reyes JA, Stoudenmire EM (2021) Multi-scale tensor network architecture for machine learning. Mach Learn Sci Technol 2:035036. https://doi.org/10.1088/2632-2153/abffe8
32. Wall ML, D'Aguanno G (2021) Tree-tensor-network classifiers for machine learning: from quantum inspired to quantum assisted. Phys Rev A 104:042408. https://doi.org/10.1103/PhysRevA.104.042408
33. Glasser I, Sweke R, Pancotti N, Eisert J, Cirac JI (2019) Expressive power of tensor-network factorizations for probabilistic modeling, with applications from hidden Markov models to quantum machine learning. In: Proceedings of NIPS, pp 1498–1510. arXiv: 1907.03741
34. Miller J, Roeder G, Bradley T-D (2021) Probabilistic graphical models and tensor networks: a hybrid framework. arXiv:2106.15666
35. Sequeira A, Santos LP, Barbosa L (2023) Soares policy gradients using variational quantum circuits. Quantum Mach Intell. https://doi.org/10.1007/s42484-023-00101-8
36. Williams RJ (2004) Simple statistical gradient-following algorithms for connectionist reinforcement learning. Mach Learn 8:229–256
37. Sanches F, Weinberg S, Ide T, Kamiya K (2021) Short quantum circuits in reinforcement learning policies for the vehicle routing problem. https://doi.org/10.48550/arXiv.2109.07498
38. Jerbi S, Gyurik C, Marshall S, Briegel HJ, Dunjko V (2021) Parametrized quantum policies for reinforcement learning. https://doi.org/10.48550/arXiv.2103.05577
39. Greensmith E, Bartlett PL, Baxter J (2004) Variance reduction techniques for gradient estimates in reinforcement learning. J Mach Learn Res 5:1471–1530
40. Schuld M (2021) Supervised quantum machine learning models are kernel methods. https://doi.org/10.48550/arXiv.2101.11020
41. Karakida R, Akaho S, Amari S-I (2019) Universal statistics of fisher information in deep neural networks: mean field approach
42. Stokes J, Izaac J, Killoran N, Carleo G (2020) Quantum natural gradient. Quantum 4:269. https://doi.org/10.22331/q-2020-05-25-269
43. Durán C, Carrasco R, Soto I, Galeas I, Azócar J, Peña V, Lara-Salazar S, Gutierrez S (2023) Quantum algorithms: applications, criteria and metrics. Complex Intell Syst. https://doi.org/10.1007/s40747-023-01073-9
44. Zhang Z, Song X, Sun X, Stojanovic V (2023) Hybrid-driven-based fuzzy secure filtering for nonlinear parabolic partial differential equation systems with cyber attacks. Int J Adapt Control Signal Process 37(2):380–398. https://doi.org/10.1002/acs.3529
45. Bhatia M, Sood SK (2020) Quantum computing-inspired network optimization for IoT applications. IEEE Internet Things J 7(6):5590–5598. https://doi.org/10.1109/JIOT.2020.2979887
46. Bhatia M, Sood SK, Kaur S (2019) Quantum-based predictive fog scheduler for IoT applications. Comput Ind 111:51–67. https://doi.org/10.1016/j.compind.2019.06.002
47. Ramezani SB, Sommers A, Manchukonda HK, Rahimi S, Amirlatifi A (2020) Machine learning algorithms in quantum computing: a survey. In: 2020 international joint conference on neural networks. IEEE, Glasgow, pp 1–8. https://doi.org/10.1109/IJCNN48605.2020.9207714
48. Cao Y et al (2019) Quantum chemistry in the age of quantum computing. Chem Rev 119:10856–10915
49. Dong Y, Zhang J (2021) An improved hybrid quantum optimization algorithm for solving nonlinear equations. Quantum Inf Process 20(4):134. https://doi.org/10.1007/s11128-021-03067-3
50. Yang J, Li B, Zhuang Z (2003) Research of quantum genetic algorithm and its application in blind source separation. J Electron 20(1):62–68. https://doi.org/10.1007/s11767-003-0089-4
51. Xiong H, Wu Z, Fan H, Li G, Jiang G (2018) Quantum rotation gate in quantum-inspired evolutionary algorithm: a review, analysis and comparison study. Swarm Evol Comput 42:43–57. https://doi.org/10.1016/j.swevo.2018.02.020
52. Lahoz-Beltra R (2016) Quantum genetic algorithms for computer scientists. Computers. https://doi.org/10.3390/computers5040024

53. Gharehchopogh FS (2022) Quantum-inspired metaheuristic algorithms: comprehensive survey and classification. Artif Intell Rev. https://doi.org/10.1007/s10462-022-10280-8
54. Gompert DC, Libicki M (2021) Towards a quantum internet: post-pandemic cyber security in a post-digital world. Survival (Lond) 63(1):113–124. https://doi.org/10.1080/00396338.2021.1881257
55. Akmal S, Jin C (2023) Near-optimal quantum algorithms for string problems. Algorithmica. https://doi.org/10.1007/s00453-022-01092-x
56. Booth KS (1980) Lexicographically least circular substrings. Inf Process Lett 10(4/5):240–242. https://doi.org/10.1016/0020-0190(80)90149-0
57. Shiloach Y (1981) Fast canonization of circular strings. J Algorithms 2(2):107–121. https://doi.org/10.1016/0196-6774(81)90013-4
58. Ambainis A (2004) Quantum query algorithms and lower bounds. In: Classical and new paradigms of computation and their complexity hierarchies. Springer, pp 15–32. https://doi.org/10.1007/978-1-4020-2776-5_2
59. Buhrman H, de Wolf R (2002) Complexity measures and decision tree complexity: a survey. Theor Comput Sci 288(1):21–43. https://doi.org/10.1016/S0304-3975(01)00144-X
60. Barenco A, Bennett CH, Cleve R, DiVincenzo DP, Margolus N, Shor P, Sleator T, Smolin JA, Weinfurter H (1995) Elementary gates for quantum computation. Phys Rev A 52:3457–3467. https://doi.org/10.1103/PhysRevA.52.3457
61. Ambainis A (2007) Quantum walk algorithm for element distinctness. SIAM J Comput 37(1):210–239. https://doi.org/10.1137/S0097539705447311
62. Vishkin U (1991) Deterministic sampling: a new technique for fast pattern matching. SIAM J Comput 20(1):22–40. https://doi.org/10.1137/0220002
63. Iliopoulos CS, Smyth WF (1992) Optimal algorithms for computing the canonical form of a circular string. Theor Comput Sci 92(1):87–105. https://doi.org/10.1016/0304-3975(92)90137-5
64. Wang Q, Ying M (2020) Quantum algorithm for lexicographically minimal string rotation. CoRR. arXiv:2012.09376
65. Childs AM, Kothari R, Kovacs-Deak M, Sundaram A, Wang D (2022) Quantum divide and conquer. CoRR. arXiv:2210.06419
66. Jeffery S (2014) Frameworks for quantum algorithms. PhD thesis. University of Waterloo. http://hdl.handle.net/10012/8710
67. Yan F, Liu W, Dong F, Hirota K (2023) A quantum-inspired online spiking neural network for time-series predictions. Nonlinear Dyn. https://doi.org/10.1007/s11071-023-08655-9
68. Hu G, Qiao C, Chen P (2021) Quantized STDP-based online-learning spiking neural network. Neural Comput Applic 33(19):12317–12332
69. Shi S, Wang Z, Cui G (2022) Quantum-inspired complex convolutional neural networks. Appl Intell 52:17912–17921
70. Welch G, Bishop G (1995) An introduction to the Kalman filter. Technical Report TR 95–041, University of North Carolina
71. Li Y, Jiang X, Zhu H, He X, Peeta S, Zheng T, Li Y (2016) Multiple measures-based chaotic time series for traffic flow prediction based on Bayesian theory. Nonlinear Dyn 85:179–194
72. Mangini S, Marruzzo A, Piantanida M, Gerace D, Bajoni D, Macchiavello C (2022) Quantum neural network autoencoder and classifier applied to an industrial case study. Quantum Mach Intell. https://doi.org/10.1007/s42484-022-00070-4
73. McClean JR, Romero J, Babbush R, Aspuru-Guzik A (2016) The theory of variational hybrid quantum-classical algorithms. New J Phys 18:2
74. Cerezo M, Arrasmith A, Babbush R, Benjamin SC, Endo S, Fujii K, McClean JR, Mitarai K, Yuan X, Cincio L, Coles PJ (2020) Variational quantum algorithms. arXiv:2012.09265 [quant-ph]
75. Bharti K, Cervera-Lierta A, Kyaw TH, Haug T, Alperin-Lea S, Anand A, Degroote M, Heimonen H, Kottmann JS, Menke T, Mok WK, Sim S, Kwek LC, Aspuru-Guzik A (2021) Noisy intermediate-scale quantum (nisq) algorithms. arXiv. 2101.08448 [quant-ph]

76. Benedetti M, Lloyd E, Sack S, Fiorentini M (2019) Parameterized quantum circuits as machine learning models. Quantum Sci Technol 4:043001
77. Mangini S, Tacchino F, Gerace D, Bajoni D, Macchiavello C (2021) Quantum computing models for artificial neural networks. EPL (Europhysics Letters) 134:10002
78. Romero J, Olson JP, Aspuru-Guzik A (2017) Quantum autoencoders for efficient compression of quantum data. Quantum Sci Technol 2(4):045001
79. Bravo-Prieto C (2021) Quantum autoencoders with enhanced data encoding. arXiv. 2010.06599 [quant-ph]
80. Lamata L, Alvarez-Rodriguez U, Martín-Guerrero JD, Sanz M, Solano E (2018) Quantum autoencoders via quantum adders with genetic algorithms. Quantum Sci Technol 4:014007
81. Khoshaman A, Vinci W, Denis B, Andriyash E, Sadeghi H, Amin MH (2018) Quantum variational autoencoder. Quantum Sci Technol 4:014001
82. Lloyd S, Schuld M, Ijaz A, Izaac J, Killoran N (2020) Quantum embeddings for machine learning. arXiv:2001.03622 [quantph]
83. Schuld M, Sweke R, Meyer JJ (2021) Effect of data encoding on the expressive power of variational quantum-machine-learning models. Phys Rev A 103:3. https://doi.org/10.1103/physreva.103.032430
84. Gil Vidal FJ, Theis DO (2020) Input redundancy for parameterized quantum circuits. Front Phys 8:297
85. LaRose R, Coyle B (2020) Robust data encodings for quantum classifiers. Phys Rev A 102(3):032420
86. Mitarai K, Negoro M, Kitagawa M, Fujii K (2018) Quantum circuit learning. Phys Rev A 98:032309
87. Lai C-Y, Chung K-M (2019) Quantum encryption and generalized Shannon impossibility. Des Codes Crypt. https://doi.org/10.1007/s10623-018-00597-3
88. Shannon C (1949) Communication theory of secrecy systems. Bell Syst Tech J 28:656–719
89. Russell A, Wang H (2006) How to fool an unbounded adversary with a short key. IEEE Trans Inf Theory 52(3):1130–1140
90. Dodis Y, Smith A (2005) Entropic security and the encryption of high entropy messages. In: Proceedings of the second international conference on theory of cryptography, series. TCC'05. Springer, Berlin, pp 556–577
91. Desrosiers SP (2009) Entropic security in quantum cryptography. Quant Inf Process 8(4):331–345
92. Desrosiers SP, Dupuis F (2010) Quantum entropic security and approximate quantum encryption. IEEE Trans Inf Theory 56(7):3455–3464
93. Bhatia M, Sood S, Sood V (2020) A novel quantum-inspired solution for high-performance energy-efficient data acquisition from IoT networks. J Ambient Intell Humaniz Comput. https://doi.org/10.1007/s12652-020-02494-x
94. Liu Y, Yang C, Jiang L, Xie S, Zhang Y (2019) Intelligent edge computing for iot-based energy management in smart cities. IEEE Netw 33(2):111–117
95. Boveiri HR, Javidan R, Khayami R (2020) An intelligent hybrid approach for task scheduling in cluster computing environments as an infrastructure for biomedical applications. Expert Syst 2020:e12536
96. Boveiri HR (2019) An enhanced cuckoo optimization algorithm for task graph scheduling in cluster-computing systems. Soft Comput 24:1–19
97. Boveiri HR, Khayami R, Elhoseny M, Gunasekaran M (2019) An efficient swarm-intelligence approach for task scheduling in cloud-based internet of things applications. J Ambient Intell Human Comput 10(9):3469–3479
98. Zhao W, Yan L, Zhang Y (2018) Geometric-constrained multi-view image matching method based on semi-global optimization. Geo-Spatial Inform Sci 21(2):115–126
99. Zhao W, Guo S, Zhou Y, Zhang J (2018) A quantum-inspired genetic algorithm-based optimization method for mobile impact test data integration. Comput-Aided Civ Infrastruct Eng 33(5):411–422

100. Liu Z, Choo K-KR, Grossschadl J (2018) Securing edge devices in the post-quantum internet of things using lattice-based cryptography. IEEE Commun Mag 56(2):158–162
101. Dey A, Bhattacharyya S, Dey S, Platos J, Snasel V (2019) Quantum-inspired bat optimization algorithm for automatic clustering of grayscale images. In: Recent trends in signal and image processing. Springer, pp 89–101
102. Park DK, Park J, Rhee J-KK (2020) Quantum-classical reinforcement learning for decoding noisy classical parity information. Quantum Mach Intell. https://doi.org/10.1007/s42484-020-00019-5
103. Cross AW, Smith G, Smolin J (2015) Quantum learning robust against noise. Phys Rev A 92:012327. https://doi.org/10.1103/PhysRevA.92.012327
104. Bernstein E, Vazirani U (1997) Quantum complexity theory. J SIAM Comput 26(5):1411. https://doi.org/10.1137/S0097539796300921
105. Ristè D, da Silva MP, Ryan CA, Cross AW, Córcoles AD, Smolin JA, Gambetta JM, Chow JM, Johnson BR (2017) Demonstration of quantum advantage in machine learning. Npj Quantum Inf 3(1):16. https://doi.org/10.1038/s41534-017-0017-3
106. Giovannetti V, Lloyd S, Maccone L (2008) Quantum random access memory. Phys Rev Lett 100:160501
107. Giovannetti V, Lloyd S, Maccone L (2008) Architectures for a quantum random access memory. Phys Rev A 78:052310. https://doi.org/10.1103/PhysRevA.78.052310
108. Hong FY, Xiang Y, Zhu ZY, Jiang LZ, Wu LN (2012) Robust quantum random access memory. Phys Rev A 86:010306. https://doi.org/10.1103/PhysRevA.86.010306
109. Arunachalam S, Gheorghiu V, Jochym-O'Connor T, Mosca M, Srinivasan PV (2015) On the robustness of bucket brigade quantum RAM. New J Phys 17:123010
110. Park DK, Petruccione F, Rhee JKK (2019) Circuit-based quantum random access memory for classical data. Sci Rep 9(1):3949. https://doi.org/10.1038/s41598-019-40439-3
111. Grilo AB, Kerenidis I, Zijlstra T (2019) Learning-with-errors problem is easy with quantum samples. Phys Rev A 99:032314. https://doi.org/10.1103/PhysRevA.99.032314
112. Gong C, Guan W, Gani A, Qi H (2022) Network attack detection scheme based on variational quantum neural network. J Supercomput. https://doi.org/10.1007/s11227-022-04542-z
113. Preskill J (2018) Quantum computing in the NISQ era and beyond. Quantum 2:79. https://doi.org/10.22331/q-2018-08-06-79
114. YanyingLiang WP, Zhu-JunZheng OS, Zhao G (2021) A hybrid quantum-classical neural network with deep residual learning. Elsevier, New York
115. Li P, Wang B (2020) Quantum neural networks model based on swap test and phase estimation. Neural Netw 130:152–164
116. Hur T, Kim L, Park DK (2022) Quantum convolutional neural network for classical data classification. Quantum Mach Intell. https://doi.org/10.1007/s42484-021-00061-x
117. Kerenidis I, Landman J, Prakash A (2019) Quantum algorithms for deep convolutional neural networks. arXiv:1911.01117
118. Liu J, Lim KH, Wood KL, Huang W, Guo C, Huang HL (2021) Hybrid quantum-classical convolutional neural networks. Sci China Phys Mech Astronomy 64(9):290311
119. Henderson M, Shakya S, Pradhan S, Cook T (2020) Quanvolutional neural networks: powering image recognition with quantum circuits. Quantum Mach Intell 2(1):2
120. Chen SYC, Wei TC, Zhang C, Yu H, Yoo S (2020) Quantum convolutional neural networks for high energy physics data analysis. arXiv:2012.12177
121. Li Y, Zhou RG, Xu R, Luo J, Hu W (2020) A quantum deep convolutional neural network for image recognition. Quantum Sci Technol 5(4):044,003. https://doi.org/10.1088/2058-9565/ab9f93
122. MacCormack I, Delaney C, Galda A, Aggarwal N, Narang P (2020) [physics, physics:quant-ph]. arXiv: 2012.14439
123. Wei S, Chen Y, Zhou Z, Long G (2021) [quant-ph]. arXiv: 2104.06918

124. Bergholm V, Izaac J, Schuld M, Gogolin C, Alam MS, Ahmed S, Arrazola JM, Blank C, Delgado A, Jahangiri S, McKiernan K, Meyer JJ, Niu Z, Száva A, Killoran N (2020) Pennylane: automatic differentiation of hybrid quantum-classical computations. arXiv:1811.04968
125. Sim S, Johnson PD, Aspuru-Guzik A (2019) Expressibility and entangling capability of parameterized quantum circuits for hybrid quantum-classical algorithms. Adv Quantum Technol 2(12):1900070

Chapter 6
Applications

Introduction by the Editor

This chapter deals with the applications of quantum computing in various fields. The applications include estimation of state of charge of batteries for electric vehicles, Kalman filter-based SoC estimation methods, developing multi-drone communication with the flying ad hoc network (FANET), degradation mechanism analysis of crystal cathode materials using machine learning, state-of-charge estimation of Lithium-ion batteries using a long short-term memory deep learning method, electrode design of multivalent metal-ion batteries, aircraft hybrid-electric propulsion, heterogeneous wireless sensor networks for obtaining a variety of sensing data sources, optimization of energy consumption in wireless sensor networks and smart home remote control security systems.

In this chapter, we, of course, provide the applications of quantum computing in electric vehicles (EVs). During the modern days, vehicles equipped with Lithium-ion battery-powered EVs are pivotal in the pursuit of decarbonisation, offering high energy efficiency, low maintenance requirements, and cost-effectiveness. Despite these advantages, accurate estimation of the state of charge (SoC) remains a significant challenge, crucial for determining remaining driving range. The QC research proposes a novel Kalman filter-based method to enhance SoC accuracy. Unlike conventional approaches, our method incorporates previous covariance matrices to address abrupt changes in battery dynamics, ensuring smoother state transitions and more precise SoC estimation. Leveraging the 2-RC Thevenin equivalent circuit model, we describe battery parameters as functions of SoC and temperature. We validate the proposed method across three distinct driving cycles, demonstrating its efficacy in minimizing SoC estimation root mean square error (RMSE) by at least 10% and delivering superior results for compact EVs.

In the era of modern technological advancements, the QC research has focused on improving network paradigms such as Flying Ad Hoc Networks (FANETs) and

V. Natarajan (ed.), *Quantum Artificial Intelligence*,
https://doi.org/10.1007/978-981-96-5051-4_6

Vehicular Ad Hoc Networks (VANETs), ushering in evolutionary ideas. Recent studies highlight the budding research in multi-drone communications, particularly with the emergence of FANETs—a swarm of Unmanned Aerial Vehicles (UAVs). FANETs offer enhanced reliability and efficacy compared to single UAV systems, making them advantageous for various operational tasks.

$LiNi_{0.5}Co_{0.2}Mn_{0.3}O_2$ (NCM523) has emerged as a prominent cathode material for lithium-ion batteries due to its high-energy density and cost-effectiveness. However, the rapid capacity fading of NCM523 poses a significant challenge to its widespread adoption. In this research, we employ scanning transmission electron microscopy (STEM) to characterize single crystal NCM523 materials under different degradation states. Subsequently, we develop and showcase a neural network model featuring a two-sequential attention block to identify crystal structures and locate defects in STEM images.

Accurate estimation of state-of-charge (SOC) is crucial for determining the driving range of electric vehicles and optimizing charge control in battery-powered systems, especially for lithium iron phosphate ($LiFePO_4$) batteries. The effects of ambient temperature and the flat form characteristics of the open circuit voltage SOC curve pose significant challenges to SOC estimation accuracy. In this research, we propose a novel SOC estimation method utilizing a Long Short-Term Memory Recurrent Neural Network (LSTM-RNN) to mitigate SOC estimation errors and model sophisticated battery behaviors under varying ambient temperatures.

Deep learning (DL) has revolutionized the prediction of materials properties for current commercial Li-ion batteries, but its application to multivalent metal-ion batteries (MIBs) is hindered by limited data availability and poor model interpretability. In this research, we develop an interpretable DL model capable of accurately learning electrode voltages for multivalent MIBs (including divalent magnesium, calcium, zinc, and trivalent aluminum) using small datasets (150–500 samples).

A research survey explores the landscape of aircraft hybrid electric propulsion (HEP), providing insights into current technologies and future trajectories across various domains including the air transport market, hybrid demonstrators, HEP topologies applications, aircraft design, electrical systems, energy storage, internal combustion engines, and management strategies. Anticipating changes in aircraft propulsion over the next three decades driven by market demand and environmental regulations, two key commercial areas are identified: electrical urban air mobility (UAM) and hybrid-electric regional aircraft. UAM is poised to enter service within the next decade, initially with small-scale devices, while hybrid-electric regional aircraft will gradually be introduced, starting with smaller aircraft as advancements in energy storage, fuel cells, aircraft design, and hybrid architectures integration progress.

Sensor nodes (SNs) in heterogeneous wireless sensor networks (HWSNs) are deployed for various sensing applications, and the mobility of sinks plays a vital role in improving system performance, energy utilization, and lifetime. Rendezvous points (RPs) are introduced to manage sink mobility, where selected SNs serve as RPs, and non-RP nodes relay data to cluster heads (CHs), which then forward the data to nearby RPs. However, determining the set of RPs and optimal paths for mobile sinks (MSs) poses significant challenges. Further, the field of self-assembly has evolved significantly from its early emphasis on creating aesthetically pleasing two- and three-dimensional structures to focusing on developing materials and devices with unique properties that were previously unattainable or prohibitively expensive. The advancements in self-assembly have facilitated the formation of even more complex structures in both two and three dimensions. The research paper "Functional Materials and Devices by Self-Assembly" describes the potential of self-assembly processes to create advanced materials and devices. Self-assembly (SA) enables the development of materials with highly ordered nanostructures that can be used in various applications such as nanoelectronics, photonics, energy storage, and chemical separations. The nanostructures help for making quantum devices.

We cover up the Water Cycle Algorithm (WCA) which has emerged as a promising bio-inspired optimization technique in the realm of Wireless Sensor Networks (WSN). In this research, we apply WCA independently to three energy models of WSN to optimize energy expenditure on data communication within the nodes. First one, Energy Reduction model: The implementation of WCA results in up to a 46.24% reduction in energy consumption in sensor nodes. Second one, Optimization Efficiency model: The research study evaluates the percentage reduction in energy depletion and the number of iterations required to achieve optimal energy usage. Third one: Simulation Insights: Different sets of node populations were analyzed, revealing notable trends in energy performance and optimization iterations.

At the end of Chap. 6, the research article "Novel Enhanced LoRaWAN Framework for Smart Home Remote Control Security" feeds an advanced security framework for smart homes using the LoRaWAN protocol. This framework or model addresses various security challenges in remote control applications within smart home environments. It protects common vulnerabilities such as impersonation and replay attacks occurring in IoT devices. It applies Elliptic Curve Cryptography to provide secure communication between devices and remote controllers. It supports energy efficiency for IoT devices in smart homes and optimizes communication protocols to reduce power consumption without compromising security. Thus, the research study concludes that with the increasing deployment of IoT devices in smart homes, ensuring robust security is important to protect user data and maintain system integrity. The innovative framework LoRaWAN illustrates significant improvements not only in securing smart home environments but also in corporate or academic quantum research labs, making it a promising solution for modern IoT applications.

Machine Generated Summaries

Disclaimer: The summaries in this chapter were generated from Springer Nature publications using extractive AI auto-summarization: An extraction-based summarizer aims to identify the most important sentences of a text using an algorithm and uses those original sentences to create the auto-summary (unlike generative AI). As the constituted sentences are machine selected, they may not fully reflect the body of the work, so we strongly advise that the original content is read and cited. The auto generated summaries were curated by the editor to meet Springer Nature publication standards. To cite this content, please refer to the original papers.

Machine generated keywords: battery, material, node, voltage, sensor, vehicle, wsn, lib, energy, lithiumion, lithiumion battery, estimation, electric, energy storage, temperature.

A Novel Method for SoC Estimation of Lithium-Ion Batteries Based on Previous Covariance Matrices and Variable ECM Parameters [1]

This is a machine-generated summary of:

Korkmaz, Mehmet: A novel method for SoC estimation of lithium-ion batteries based on previous covariance matrices and variable ECM parameters [1].

Published in: Electrical Engineering (2022).

Link to original: https://doi.org/10.1007/s00202-022-01692-4

If you want to cite the papers, please refer to the original.

For technical reasons we could not place the page where the original quote is coming from.

Abstract-Summary

"Accurate estimation of state of charge (SoC) of batteries, which refers to the remaining driving range, is one of the most notable challenges for EVs."

"In this paper, a novel Kalman filter-based estimation method is proposed to increase the accuracy of the SoC. The proposed method considers not only the current but also the previous covariance matrices since abrupt changes in the nonlinear dynamics of the battery may lead to incorrect estimation."

"Smoother state transitions are provided, and more accurate SoC estimation is possible."

"The improved method is supported by the 2-RC Thevenin equivalent circuit model, whose parameters are described as a function of the SoC and temperature."

"The battery model and proposed method are tested with three different driving cycles to prove the efficiency."

"The proposed method can minimize the RMSE of SoC estimation up to at least 10% and provides better SoC estimations for compact EVs."

Introduction

"The latter, however, is more preferred for the SoC estimation in practice, inasmuch as the OCV approach requires an extended rest time and cutting off the power, the CC method contains cumulative errors and initial values must be known, and the electrochemical equations cannot be modeled accurately since the equations have temperature sensitivity, features of batteries and time-dependent nonlinear partial differential equations, which have to be calculated every time."

"As NN-based methods, heuristic optimization algorithms are also a widely used and frequently preferred tool when estimating the SoC since they find optimal ECM parameters that can characterize steady and transient states of the LiBs well."

"In a nutshell, the main contribution of the study at hand can be summarized as follows: It updates the ECM parameters at each step of a driving cycle according to the environment temperature and current SoC change so that it yields better SoC estimations since battery characterization can be defined well."

Battery Equivalent Electrical Circuit Model and Parameter Identification

"Equivalent electrical circuit models (ECM) are often used to model the internal electrochemical reactions of lithium-ion batteries during charge/discharge regimes."

"It ignores the temperature fluctuations, the current SoC value, and electrolyte concentration and has therefore been seen as an unsuccessful representation in ECM models."

"Better ECMs can be achieved by adding parallel RC branches to the R_{int} model and thus, instantaneous and delayed system responses can be modeled well."

"2-RC Thevenin ECM is a widely accepted choice by researchers as it provides a sufficient level of accuracy in battery modeling In parallel with the literature, it is also used 2-RC Thevenin ECM in this study [2]."

"Determining the different circuit parameters in ECM plays a pivotal role in the accurate estimations of the SoC. In literature, several ECM parameters can be found from heuristic approaches to model identification."

Kalman Filter-Based SoC Estimation Methods

"As the Kalman filter produces optimal results for linear systems, it is often used for state estimation of such systems."

"Just as many engineering systems, batteries also exhibit a strong nonlinear structure due to electrochemical effects, environmental effects, etc. To overcome this issue, the EKF structure, which linearizes the existing mean and covariance values of the systems, is frequently employed instead of classical KF to estimate the battery SoC. Algorithm 2 summarizes SoC estimation using the EKF algorithm."

"Instead of EKF, researchers proposed a UT transform-based KF, namely UKF, for the estimation of nonlinear systems."

"Similar to EKF, state estimation is performed using recursive prediction and correction steps (Algorithm 3)."

"When this situation is taken into account, it is thought that choosing sigma points by taking into account the covariance matrices of the previous cases may produce a better and more robust state estimation."

Results

"In order to test the acquired battery model, it is benefited from LA92, UDDS, and a combination of LA92, UDDS, US06, and The Highway Fuel Economy Test (HWFET) driving cycles."

"LA92 driving cycle experiment was conducted to evaluate the battery model and its results are compared for the algorithms."

"Analyzing the Figure, SoC estimation results of all algorithms closely tracked the reference SoC curve, as did in LA92 test, but the best one was acquired with the proposed method."

"That being median values much and distribution box size large compared to the LA92 test is probably related to the more total running time of UDDS than LA92."

"To this intent, to verify the proposed algorithm's effectiveness, a combination of LA92, UDDS, HWFET, and US06 cycles was applied to the battery model."

Conclusion

"Based on the battery model and driving cycles, EKF, UKF, and SRUKF algorithms were implemented along with the proposed method to compare the results."

"The proposed algorithm, PoCSRUKF, has better estimates rather than previous approaches for all driving cycles."

"The median of SoC error is at least 20% better than other methods for all driving cycles, and especially for the LA92 driving cycle, it has almost 7 times lower."

"RMSE values for all different driving cycles and algorithms confirm the superior performance of the proposed method."

Generic Potential Field Based Distributed Node Coordination in Flying Adhoc Network (FANET) [3]

This is a machine-generated summary of:

Meena, T.; Sangam, Ravi Sankar: Generic potential field based distributed node coordination in flying adhoc network (FANET) [3].

Published in: Journal of Ambient Intelligence and Humanized Computing (2022).

Link to original: https://doi.org/10.1007/s12652-022-03767-3

If you want to cite the papers, please refer to the original.

For technical reasons we could not place the page where the original quote is coming from.

Abstract-Summary

"Latest studies have shown that, the research is budding in multi-drone communications with the swarm of UAVs called flying adhoc network (FANET)."

"Not only are FANETs more reliable than single UAV but are also likely to be more efficacious and advantageous in finishing operational tasks."

"Due to the high mobility characteristic, FANETs don't have a fixed topology and expeditiously changes its topological structure which makes coordination between UAVs in the FANET arduous."

"The extant solutions used in the conventional adhoc network for node coordination cannot be applied in the FANET."

"This paper addresses the above mentioned issues by proposing a novel distributed node coordination algorithm for FANETs."

"The crux in this proposed algorithm is to make use of the General Potential Field based node coordination (GPFnc)."

"The results of the experiment clearly demonstrated that the proposed GPFnc achieves enhanced scalability, reliability and fast network formation than the existing contemporary algorithms of MANET."

Introduction

"This seems to be more reliable in FANET, However these protocols are also having some issues like link disconnections, security metrics, and fault tolerance, optimal path planning etc. Out of these issues, node coordination plays a crucial role in FANET because coordinating the other UAV can make the network more reliable and robust."

"The contributions in this paper are as follows: To analyze Distributed Node Coordination in Flying Adhoc Network using Generic Potential Field Validation of the proposed algorithm through the simulation and examining the navigation of UAVs Performance analysis on every metrics of UAV and FANET including battery lifetime, resource usage and utilization etc. Performance comparison and validation of the proposed GPFnc algorithm with the state of the art algorithms."

"As per our knowledge, this is the first unique algorithm to utilize Generic Potential Field for node (UAV) coordination in FANET."

Literature Survey

"We tried to address the issues of node coordination in the MANET and we also incorporates few routing, path planning algorithms in the literature for Multi-UAVs."

"Khan et al. [4] developed a self-organization based clustering scheme by studying the behaviors of glowworm swarm optimization (GSO) which is applicable to UAVs in the Flying ad-hoc Network (FANET)."

"Distributed coverage path planning, route topology discovery and finding optimal UAV relay deployment were the key parts implemented in the algorithm."

"Routing algorithm based on 3D graph theory was designed by Razzaq et al. [5] Collision free path planning was provided by the router for UAV applications."

"To improve efficiency, a network based solution needs to be proposed by integrating UAV self-coordinating autonomy algorithms with routing schemes."

"Various cluster-based routing protocols available for MANET can suffer with the parameters such as cluster mobility, energy efficiency, data transmission, and end-to-end delay etc., when deployed in the FANET (Radmanesh et al. [6]; Anicho et al. [7])."

Generic Potential Field

"The scalar potential of the gravitational field (Li and Du [8]) is expressed as $\Phi(X) = GM/r$, where X is the spherical coordinates with mass point M, the gravitational potential at a distance r and the gravitational constant G. If n mass points are present in the space, the total potential at point X is just the algebraic sum of the potentials due to the individual mass point is expressed as: From the above mention potential field, the potential energy at any point in space is directly proportional to the strength of the object interaction."

"The potential is the conventional field which is directly proportional to the mass and inversely proportional to the particle distance."

"Agreeing the supervision standard of the potential data field, the above nuclear like and gravitational like potential function can be attained as, The proven difference between the properties of probability density and potential data field functions are normalize constant if K is the finite in space."

Experimental Setup and Performance Analysis

"To prove the efficacy of the proposed GPFnc, the resource monitoring is carried out for the two scenarios."

"The performance validation is carried in various scenario leading the pathway to prove the proposed algorithm."

"Further, the proposed GPFnc is compared with the state of the art algorithms available in the MANET in order to prove the efficacy of the algorithm."

"Further it is noted that the proposed GPFnc is lesser in terms of battery utilization and the evaluation of this metric is carried out along with the onboard processing of other sensors."

"The same evaluation is also carried out for the other state of the art algorithms in order to prove the efficacy of the proposed GPFnc."

"The same setup is used to validate the proposed GPFnc and the state of the art algorithms."

Conclusion

"The authors of this paper have pioneered a novel algorithm—Generic Potential Field based node coordination algorithm (GPFnc) for FANET."

"In line with previous studies, we believe this is the first work to utilize Generic Potential Field in node coordination and the simulation evinces that GPFnc algorithm works effectively in node coordination."

"On comparing the experimental results with other sophisticated algorithms in MANET, it was observed that the results showcased by GPFnc were more accurate."

Degradation Mechanism Analysis of $LiNi_{0.5}Co_{0.2}Mn_{0.3}O_2$ Single Crystal Cathode Materials through Machine Learning [9]

This is a machine-generated summary of:

Sha, Wuxin; Guo, Yaqing; Cheng, Danpeng; Han, Qigao; Lou, Ping; Guan, Minyuan; Tang, Shun; Zhang, Xinfang; Lu, Songfeng; Cheng, Shijie; Cao, Yuan-Cheng: Degradation mechanism analysis of $LiNi_{0.5}Co_{0.2}Mn_{0.3}O_2$ single crystal cathode materials through machine learning [9].

Published in: npj Computational Materials (2022).

Link to original: https://doi.org/10.1038/s41524-022-00905-5

If you want to cite the papers, please refer to the original.

For technical reasons we could not place the page where the original quote is coming from.

Abstract-Summary

"The rapid capacity fading of NCM severely hinders its development and applications."

"The single crystal NCM523 materials under different degradation states are characterized using scanning transmission electron microscopy (STEM)."

"The number of point defects in NCM523 is observed to experience a trend of increasing first and then decreasing in the degradation process."

"This analysis sheds light on the defect evolution and chemical transformation correlated with layered material degradation."

Introduction

"The advanced electron microscopy investigations on the degradation mechanisms of NCM play indispensable roles in the design of high-energy-density lithium-ion battery materials."

"Current electron microscopy allows the direct visual recognition of atom-level configuration information from the host crystal structure, lattice plane distortion to even single point defect, providing rich information of battery material processes and properties."

"It is urgent to develop an automatic framework for material characterization image processing, which can efficiently locate and track atom defects, or other microstructure configurations, and quantitatively study the property evolutions in materials."

"A deep learning u-net network, one of the common AI algorithm frameworks, is improved to study the capacity degradation mechanism of ternary cathode materials."

"This research consists of sample preparation of ternary materials, electrochemical and electron microscopy characterization, and machine learning."

Results and Discussion

"The atomic simulation environment, a Python library for working with atoms, is employed to generate a large number of atomic cluster models, including layered, spinel, and rock-salt structures."

"Transition metal vacancies, lithium vacancies, interstitial atoms, and other point defects are introduced into the primary crystals of these atomic models."

"The input of the network is the simulated STEM images of different atom models."

"The proposed u-net model is employed to locate the atomic defects in the NCM STEM experimental images."

"The point defect ratio is the ratio of defect numbers to all atoms in the STEM images."

"Through the high-throughput deep learning of NCM electron microscopy data, this research classifies and counts the states, categories, and quantity of material crystal defects, explores the corresponding relationship between cycle life and microscopic crystal structures, and provides more profound insights into the aging mechanism and physical and chemical reaction of ternary layered cathode materials."

Methods

"NCM523 (Sigma-Aldrich) powders as cathode materials and graphite as anode materials were purchased to assemble pouch cells."

"The graphite slurry with a weight ratio of graphite:binder:SP = 8:1:1 were coated on Cu foils as the anodes."

"The NCM slurry with a weight ratio of NCM523:binder:SP = 7:2:1 were coated on Al foils to form the cathodes [10]. M LiPF6 solution was selected as the electrolyte, and the solvent was a mixture of vinyl carbonate (EC), dimethyl carbonate neutralization (DMC), and methyl ethyl carbonate (EMC) in a volume ratio of 1:1:1."

"The cathode films were obtained by disassembling the pouch cells and wetting the cathode surfaces with a few drops of DMC solvent."

"The cathode materials were scraped from the aluminum foil and immersed in N-methyl-2-pyrrolidone solvent to remove the residual PVDF binders."

State-of-Charge Estimation of Lithium-ion Batteries Using LSTM Deep Learning Method [11]

This is a machine-generated summary of:

Chung, Dae-Won; Ko, Jae-Ha; Yoon, Keun-Young: State-of-Charge Estimation of Lithium-ion Batteries Using LSTM Deep Learning Method [11].

Published in: Journal of Electrical Engineering & Technology (2022).
Link to original: https://doi.org/10.1007/s42835-021-00954-8

If you want to cite the papers, please refer to the original.
For technical reasons we could not place the page where the original quote is coming from.

Abstract-Summary
"We proposed a SOC estimation method by using a long short-term memory (LSTM)–recurrent neural network (RNN) to reduce the SOC estimation errors, and to develop a model for the sophisticated battery behaviors under varying ambient temperatures, including time-variable current, voltage, and temperature conditions."

"The experimental results show that the proposed method can accurately learn the influence of ambient temperatures on the battery and also estimate the battery's SOC under varying temperatures with root mean square errors less than 1.5% and mean average errors less than 1%."

"The main contribution of this study is the comprehensive explanation and implementation process of the data-based DL approach for the SOC estimation of the LIBs in the following aspects, (1) An LSTM-RNN was trained to model the complex battery dynamics under varying ambient temperatures. (2) The proposed method is model-free and data-driven approach, which means there is no need to construct OCV-SOC lookup tables under varying temperatures in order to pick an appropriate equivalent circuit model."

"The proposed method can be extended for the SOC estimation of other types of lithium batteries."

Introduction

"It is still a challenging work to determine the SOC from measurable quantities such as terminal voltage, charging and discharging current, and ambient temperature because of the highly sophisticated properties of battery dynamics, along with unpredictable conditions, such as regeneration, hysteresis, and aging [12]."

"To get over these problems, an LSTM-RNN was also proposed in this work for the complex battery internal dynamics under varying temperatures and to estimate the battery's SOC from the voltage, current, and temperature variables."

"The main contribution of this study is the comprehensive practical approach and application process of the data-based DL approach for the SOC estimation of the LIBs in the following aspects: (1) An LSTM-RNN was applied to train the model of the complex battery dynamics under varying ambient temperatures conditions."

"The network sufficiently estimates the battery's dynamics with presenting good robustness against unknown parameters including initial states, and provides an accurate SOC measurement with varying temperatures conditions. (2) The proposed method is the model-free and data-driven approach."

Lithium-ion Battery Characteristics

"The LIB has been payed the attention of R&D groups and automobile industries because of its high energy density, low self-discharge rate, high cell voltage, long lifespan, low weight, and fast recharging characteristics [13] compared with classical lead acid batteries."

"An accurate measurement of the temperature inside the LIBs and an understanding of the temperature effects are important for proper battery management."

"The performance comparative analysis between the LIB and other EV batteries in terms of nominal voltage, life cycle, depth of discharge, and efficiency demonstrates that the LIB appears to be a better choice for the EV application."

"An accurate SOC measurement enables the LIB to disconnect by cutting-off the charging and discharging circuitry in the battery management system (BMS) whenever it is out of the safe operating conditions, and to trigger battery charging under normal operating conditions."

Long Short-Term Memory (LSTM) Model for SOC Estimation

"The NNs have also efficient in time-sequenced data processing because of their parallelism in computation."

"The temporal relations in time-sequence measured data are the key for modeling sequential data, such as natural language processing, time-dependent systems like battery SOC estimation problems."

"Unlike feedforward NNs, which totally ignore the past input data, the RNN [14] stores historically past information with recurrent neuron units."

"As the RNN naturally encodes historical input data, it is superior at tasks with sequential input such as handwriting recognition and vocal translation in time-sequence measured data."

"The classic RNN has a limit of not handling long-term time dependency, as the influence of past inputs fade out exponentially with time [15]."

"The errors can also flow backwards for many temporal steps data, which prevents the gradient vanishing problem and enhances the network's capability of capturing long-term dependencies."

Dataset Collection and LSTM Training Setup for Deep Learning

"A forward pass computation starts when the training data are fed into the networks and ends when the SOC estimation are generated at each time step as well as when the errors and the overall loss are calculated."

"An LSTM network is a supervised learning algorithm that learns a function by training on a given set of features and a target output data, As the supervised learning is a deep learning task of learning a function that maps an input to an output based on the desired input–output pairs to make predictions for future outcomes, it determines a function from the labeled training data for the desired output values."

"The LSTM network takes the battery current, terminal voltage, and ambient temperature as the network input data variables, while it takes the SOC as the output variables at the same cycle."

LSTM Deep Learning Results and Discussion

"In order to investigate the effects of ambient temperature, the proposed network was trained to estimate the battery SOC under varying temperatures."

"Because the network was also trained with data at room temperature, the proposed LSTM network yielded a satisfying SOC estimation, as shown in panel (b) of each figure."

"The network presents a better estimation at low temperatures (0–10 °C), particularly during the 80%–40% SOC, as at shown in panel (b) of each figure."

"For the SOC estimation at room temperature, the corresponding RMSE and MAE are 1.06% and 0.93%, respectively."

"The LSTM network yields a quite satisfying SOC estimation, with RMSEs and MAEs less than 2.1% and 1.6%, respectively."

"Considering the poor estimation under low temperatures, one practical solution to improve the SOC estimation is to train networks particularly after obtained the testing data for batteries working at low temperatures."

Conclusions

"An SOC estimation approach was proposed to apply an LSTM neural network."

"To estimate the battery's SOC more accurately, the measured voltage, current, and temperature were taken as the inputs of the neural network, while the output was the SOC."

"An LSTM network was proposed to train the complex battery internal model under varying ambient temperatures."

"The network was well trained the battery dynamics, presents good performance against unknown battery parameters including initial states, and provides satisfying SOC estimations under varying temperature conditions."

"The proposed method need not to consider to build an OCV-SOC lookup table, unlike analytic equivalent circuit model-based SOC estimation."

"This method does not need the physics of the batteries, since the neural network is a data-driven approach."

Interpretable Learning of Voltage for Electrode Design of Multivalent Metal-Ion Batteries [16]

This is a machine-generated summary of:

Zhang, Xiuying; Zhou, Jun; Lu, Jing; Shen, Lei: Interpretable learning of voltage for electrode design of multivalent metal-ion batteries [16].

Published in: npj Computational Materials (2022).

Link to original: https://doi.org/10.1038/s41524-022-00858-9

If you want to cite the papers, please refer to the original.

For technical reasons we could not place the page where the original quote is coming from.

Abstract-Summary

"Deep learning (DL) has indeed emerged as a powerful tool for rapidly and accurately predicting materials properties from big data, such as the design of current commercial Li-ion batteries."

"Its practical utility for multivalent metal-ion batteries (MIBs), the most promising future solution of large-scale energy storage, is limited due to scarce MIB data availability and poor DL model interpretability."

"We develop an interpretable DL model as an effective and accurate method for learning electrode voltages of multivalent MIBs (divalent magnesium, calcium, zinc, and trivalent aluminum) at small dataset limits (150–500)."

Introduction

"The reported mean absolute error (MAE) of the deep-neural network (DNN) model is lower than conventional ML in predicting the volume change and voltage of LIB electrodes [17, 18]."

"Joshi and others predicted the voltages of Na-ion battery electrodes with a DNN model, but MAE is much higher than that of LIBs [17]."

"The high-performance and interpretable DL model is of high need for predicting a variety of properties of multivalent MIBs from small data and then designing high-performing multivalent MIBs."

"We firstly train our DL models with relatively large data of the electrode voltage of LIBs (2000+ data) from Materials Project (MP) [19, 20]."

"Our results show that the highly accurate and interpretable deep model could accelerate the discovery and design of electrode materials for multivalent MIBs and the development of the large-scale battery industry."

Results and Discussion

"After fully training, MAE of our DL model for predicting the voltages of LIBs is only 0.32 V, which is lower than the conventional ML results (0.40 V)."

"The DL model, CGCNN, and the two conventional ML models are not only used for the Li-ion battery voltage prediction, but also the multivalent MIBs, while the TL model is only used for the multivalent MIBs."

"Our model outperforms the conventional ML model in predicting the electrode voltage for multivalent Mg-, Ca-, Zn-, and Al-ion batteries."

"We develop an interpretable TL model to accurately predict the electrode voltages for MIBs (especially the multivalent batteries with very small data) and explain the underlying physical pictures as the important features for the voltage prediction by visualizing the vectors in layers in the neural network."

Methods

"The CGCNN model presents a periodic crystal structure into a multigraph G. Each atom in a structure is represented by a node i in G, which is represented by the atomic feature vector v_i."

"In the CGCNN model, only elemental and structural features of the electrode crystals are used, because our training datasets retrieved from Materials Project database are on the basis of DFT calculations."

"In the CGCNN network [21], the output vectors of each layer are available for public, the ante-hoc could be used to probe the whole model."

"Local voltage representation is derived from the local oxygen-coordination environment vectors and represents the contributions of each atom to the voltages of the crystal."

"The local voltage of elements in the oxygen-coordination environment is a specific local voltage for the atom, having at least two nearest neighboring oxygen atoms in the crystal."

Aircraft Hybrid-Electric Propulsion: Development Trends, Challenges and Opportunities [22]

This is a machine-generated summary of:

Rendón, Manuel A.; Sánchez R., Carlos D.; Gallo M., Josselyn; Anzai, Alexandre H.: Aircraft Hybrid-Electric Propulsion: Development Trends, Challenges and Opportunities [22].

Published in: Journal of Control, Automation and Electrical Systems (2021).

Link to original: https://doi.org/10.1007/s40313-021-00740-x

If you want to cite the papers, please refer to the original.

For technical reasons we could not place the page where the original quote is coming from.

Abstract-Summary

"The present work is a survey on aircraft hybrid electric propulsion (HEP) that aims to present state-of-the-art technologies and future tendencies in the following areas: air transport market, hybrid demonstrators, HEP topologies applications, aircraft design, electrical systems for aircraft, energy storage, aircraft internal combustion engines, and management and control strategies."

"Two commercial areas are in evolution, electrical urban air mobility (UAM) and hybrid-electric regional aircraft."

"The last one will gradually come into service, starting with small aircraft according to developments in energy storage, fuel cells, aircraft design and hybrid architectures integration."

"Turbo-electric hybrid architecture combined with distributed propulsion and boundary layer ingestion seems to have more success for regional aircraft, attaining environmental goals for 2030 and 2050."

"Computational models supported by powerful simulation tools will be a key to support research and aircraft HEP design in the coming years."

Introduction

"Whereas air transport industry is responsible for a considerable part of greenhouse gas emissions, a concept where ICEs and electric motors (EM) are combined in the propulsion to increase vehicle efficiency and reduce the impact is being analyzed (Zhang et al. [23])."

"Some of the main advantages of HEP compared with the traditional propulsion are: (a) increasing the global aircraft efficiency; (b) increasing aircraft reliability, power distribution/quality, and flight range; (c) emissions and noise reduction; (d) capacity of extending the market to smaller airports (Sliwinski et al. [24])."

"Hybrid-electric propulsion system (HEPS) appears as the most viable solution for an energy efficient, cleaner and quieter aeronautical propulsion, since it is able to combine the advantages of the conventional propulsion system and the all-electric approach (Sliwinski et al. [24])."

"Recent advances in electrical motors, energy storage systems, and power electronics converters (PEC) are leading the aircraft propulsion to become increasingly electrical (Sliwinski et al. [24])."

Civil Aviation Market

"Other programs such as CLEEN in the USA, GARDN in Canada, Clean Sky and Clean Aviation in the European Union (EU) focus in developing certifiable aircraft technology for fuel consumption reduction, silent flying, and reduced emissions (IATA [25])."

"These new aircraft concepts with architectures that increase propelling torque and power, for shorter takeoff distances, can enable future devices to utilize smaller airports (Liu [26])."

"Airports as Schiphol in Amsterdam encourage the use of quiet and clean aircraft, while flag that infrastructure adjustments with additional energy supply might be required for electric aircraft, for battery storage and shipping."

"Aircraft electric propulsion (EP) will influence the aviation business on the airports."

"The expected penetration of EP aircraft into the market would start with 1–2 passengers all-electric urban air taxis (UAT) until 2025, 15–20 passengers HEP aircraft from 2025 to 2030, and battery or FC powered aircraft from 2035 to 2045 (IATA [25])."

Aircraft Hybrid-Electric Propulsion

"Battery-powered EM and ICE shafts are both coupled to a shaft that drives a fan or propeller, so either or both can provide propulsion (National Academies of Sciences, Engineering and Medicine [27])."

"Some concepts are testing a variation of this architecture combined with DP, in which one or more fans are mechanically driven by ICEs, while others exclusively by EMs, which are powered by a battery package or by the ICE-driven electrical generator (National Academies of Sciences, Engineering and Medicine [27])."

"This architecture is well suited to combine with DP concepts, where the power is distributed to several EM-driven fans, positioned to provide synergistic integration in the airframe, and also with superconducting electrical systems."

"Since fans usually operate at lower speed than common EMs, turbo-electric and series hybrid architectures may drive geared turbofan engines that reduce shaft speed, with the benefit of shaft speed decoupling from ICE (Brelje and Martins [28])."

Developments in Hybrid-Electric Propulsion

"The small airliner with one 500 kW ICE-driven electric generator, powers two EM-driven fans with a cruising speed of 483 km/h, declared range of 1110 km, takeoff distance of 670 m, and bays on wings for modular battery packs."

"The Parallel Electric-Gas Architecture with Synergistic Utilization Scheme (PEGASUS) is a 48-passenger regional aircraft concept based on the ATR 42–500, and holds a TRL of 4–5 (HECARRUS [29])."

"The Wright 1 is a 186-passenger all-electric aircraft concept developed in partnership with the airline EasyJet, with a TRL of 3–4 (HECARRUS [29]), and projected to fly on 2035 (Wright Electric [30]; Misra [31]; IATA [25]). The Bauhaus Luftfahrt concept Ce-Liner is a 189-passenger and 1670 km autonomy aircraft, with C-wing configuration, and envisaged EIS by 2035 (IATA [25]), also with a TRL of 3–4 (HECARRUS [29])."

"It is a 3400 kg aircraft that employs a Jet-A fueled piston ICE, a hybrid-electric system in series architecture, and several EM-driven propellers in DP (Continental [32]; EEPower [33])."

"The NASA Scalable Convergent Electric Propulsion Technology and Operations (SCEPTOR) is a flight demonstrator project that retrofit the ICE powered Tecnam P2006T aircraft, removing the wing and replacing it with an experimental wing integrated with EMs (Borer et al. [34]; Patterson et al. [35])."

Developments in Aircraft Design and Systems Integration

"The aircraft design starts with the requirements and specifications, and later the definition of constraints derived from airworthiness, aerodynamics, safety, etc. Propulsion system specification on HEP must evaluate the engines, ignition system, fuel supply, cooling control, exhaust management, etc. (González Espasandín et al. [36])."

"The conceptual design is based on aerodynamic, propulsion and structural solutions, and utilizes user-friendly software such as Advanced Aircraft Analysis (AAA) by DAR, RDSWin, Simulating Aircraft Stability and Control Characteristics

(SimSAC), and Stanford University Aerospace Vehicle Environment (SUAVE), this last developed in collaboration with EMBRAER (Sziroczak et al. [37])."

"Novel aircraft configurations such as hybrid wing body (HWB) (Kim et al. [38]), blended wing body (BWB) (Yang et al. [39]; Boeing [40]; IATA [25]), strut-braced wings (SBW) (Harrison et al. [41]; Bradley and Droney [42]), Fixed Wing (FW) (Van Oppen et al. [43]) Transonic Truss-Braced Wing (TTBW) (Boeing [44]), and Box-Wing Tasca [45]; Palaia [46] are more environmentally friendly and quieter than conventional designs."

"A design trend on hybrid-electric architectures is to employ the EMs to deliver the additional thrust during takeoff, while the ICE could be reduced in size (Sliwinski et al. [24])."

Developments in Aircraft Electrical Systems

"High-conversion high-voltage (HCHV) DC-DC converters with higher DC bus voltages are a trend in electric aircraft development, aiming to reduce size and weight and to increase power density."

"EMs with high power-to-weight ratios (PWR) are required for HEP applications (Corduan et al. [47])."

"High specific power density EM is a key technology for HEP, and high fundamental frequency combined with genetic algorithms is an approach for air core machines design (Zhang et al. [23])."

"Li-S battery is a promising technology for aircraft use, with a theoretical specific energy of 2600 Wh/kg, suitable to operate at very low temperatures, and to be installed in thin-film layers along the wing distributing the weight and reduced cost of sulfur (Gao et al. [48]; Sliwinski et al. [24])."

"PEMFC fueled by hydrogen have a lower weight than solid oxide FCs (SOFC), and work in low temperature, high current densities, quick start-up, high energy density (600–800 Wh/kg) and low power density (0.5–1 kW/kg)."

Developments in Aircraft Internal Combustion Engines

"Jet engines are classified in turbojet (for combat aircraft), turbofan (for large commercial aircraft), or turbo-prop (for medium aircraft) (González Espasandín et al. [36])."

"Safran works on the Ultra-High-Bypass Ratio (UHBR) design, with extensive use of composite materials to achieve a BPR of 15, and an expected reduction of up to 25% in fuel burn relative to conventional engines (Vetters et al. [49])."

"General Electric Catalyst is a turbo-prop ICE that claims to explore the full range of pitch using a Full Authority Digital Engine Control (FADEC) with integrated propeller control for an improvement of 15% in fuel consumption (GE Aviation [50])."

"It is a four-stroke rotary combustion ICE, with vantages such as simple structure, multi-fuel, higher speed, high power density, low noise and low vibration (Chen et al. [51]; Yang et al. [52])."

Developments in Flight Control and Energy Management

"In Bongermino and others [53], a complete model of an UAVE on parallel architecture HEP was developed, and a control strategy for energy management and optimization of the propulsion system was developed."

"An optimal energy management problem is analyzed for a propulsion system consisting of an ICE and a battery-powered EM in HEP on parallel architecture."

"A HEP control with an artificial neural network is described presenting the results of an optimization strategy, and minimizes the power consumption of the ICE from the electrical source on an UAVE."

"A sensitivity analysis is performed in a HEPS for a defined aircraft in series architecture, to study the influence of variations on specific power and efficiencies of electric components (PECs and EMs), on the MTOW and fuel burn."

Conclusions

"The most critical aspects for aircraft industry in the next years are sustainable growth, low operational costs and look for opportunities coming from the new technologies."

"To attain environmental goals, all-electrical and hybrid-electrical propulsion will modify the aircraft market next years."

"All-electrical seems to be well suited for UAVEs, while turbo-electric is the base of the most cited HEP concepts for medium size aircraft, in concepts that combine it with DP and BLI."

"Series architecture seems to be applicable for small aircraft combined with FC or Wankel ICEs."

"Various HEP demonstrators have been developed, most of them on small-size aircraft of up to 4 passengers, with series or parallel architectures, and powered by ICEs or FCs."

"PECs are testing different topologies and control strategies, aiming to deal with challenging demands of aircraft environment such as HV circuits, fast changing operation conditions of EM-driven propellers, low pressure environments and TMS."

Energy Efficient Rendezvous Points Based Routing Technique Using Multiple Mobile Sink in Heterogeneous Wireless Sensor Networks [54]

This is a machine-generated summary of:

Gupta, Preeti; Tripathi, Sachin; Singh, Samayveer: Energy efficient rendezvous points based routing technique using multiple mobile sink in heterogeneous wireless sensor networks [54].

Published in: Wireless Networks (2021).

Link to original: https://doi.org/10.1007/s11276-021-02714-y

If you want to cite the papers, please refer to the original.

For technical reasons we could not place the page where the original quote is coming from.

Abstract-Summary

"Sensor nodes (SNs) are used in numerous uses of heterogeneous wireless sensor networks (HWSNs) to obtain a variety of sensing data sources."

"Sink mobility shows a significant part in the enhancement of sensor system execution, energy utilization, and lifetime."

"To manage sink mobility, rendezvous points (RPs) are introduced where some SNs are chosen as RPs, and the non-RP nodes convey the information to the cluster heads (CHs)."

"To determine the set of RPs and travelling path of mobile sinks (MSs) that visits these RPs is quite challenging."

"This work presents an energy-efficient SOSS based routing method that depends on RPs and multiple MSs in HWSNs."

"The use of multiple MSs can enhance the data collection efficiency and decreases the energy utilization for HWSNs."

"The hybrid seagull optimization and salp swarm (SOSS) algorithm is used to find the RPs and travelling routes of MS."

Introduction

"The usage of MSs can significantly enhance the energy efficiency of SNs in WSNs than static sinks [55]."

"During data communication and reception, to reduce the energy utilization, CHs are utilized to execute the routing among CMs and sink nodes."

"The subsets of SNs are chosen as RPs, and non-RPs nodes transfer the data to the RPs [56]. In multi-hop communication, the issue which is termed rendezvous design selects appropriate RPs to reduce the utilization of energy while a given packet delivery bound."

"To enhance the energy efficiency by mobile sink based HWSNs data gathering, and also it propose a dynamic mobile sink path selection for improving network lifetime in HWSN."

"The usage of multiple MSs can enhance the data collection productivity, expand the lifespan of the network, and lessen the utilization of energy in HWSNs."

Related Works

"To produce a large amount of energy efficient steering trees and to balance the node vitality utilization, a new rotation heuristic is merged to the top-down CH routing tree."

"By computing the weighted determination likelihoods of every SNs, E-BEENISH procedure elects the CHs, which depends on the distance from the MS to the node and remaining energy."

"An energy efficient routing approach and a novel relay node placement was developed by Zhang and others [57] in HWSNs."

"The key goal of the work was to solve both issues of energy efficient routing and relay node placement (RNP) for HWSNs."

"The system considered that the HWSN consist of unreachable region for RNP issue and converted for an energy efficient routing to route distance of WSN."

"A SERA (Stable Energy Efficient Routing algorithm) was presented by Sunil and others [58] for dynamic HWSNs."

Proposed Methodology

"Energy effective RP-based routing procedure is developed using multiple MS in HWSNs."

"In the proposed system, the WSN contains heterogeneous nodes and has distinct levels of nodal energy."

"The suggested scheme comprises three-stage like cluster creation, selection of CH, MS routing."

"Multiple MSs are utilized in the routing process."

"The selection of RPs is the key problem in MS-based routing."

"Each MS can move within the network region and visits certain RPs for sensory data collection."

"To solve the above-mentioned problems, we have developed a SOSS protocol by incorporating SOA and SSA to discover an optimum set of RPs and travelling routes of the MS."

"In the network, the heterogeneous nodes are arbitrarily spread and all the nodes have different processing capabilities."

"Super nodes are minimum energy related to the advanced nodes and greater than normal nodes."

Proposed Work

"Eagle chooses a place with the maximum number of prey in the initial phase."

"Each CH must be covered by at least one RP, Therefore, by reducing the total number of RPs, the entire path length is decreased for the MS."

"The key point of creating a hybrid procedure is to find an ideal set of RPs and travel routes of the MS."

"The hybrid SOSS procedure is initiated by the input parameters such as the number of MSs, number of CMs, the quantity of CHs, amount of RPs, and the extreme quantity of iterations."

"The working procedure of hybrid SOSS is stated below: Using SOA, each MS can travel in the sensing area for routing."

"The minimum number of RPs is the fitness function for the hybrid SOSS algorithm."

"The fitness function for the proposed hybrid SOSS algorithm is given as: SSA procedure is utilized to update the MSs Path."

Results and Discussions

"The proposed SOSS is related to the existing methods such as, ETSSEP, DETSSEP [59], TEDRP, ECDA, EHDT, and NFECG [60]. The performance is evaluated in terms of dead nodes, alive nodes, throughput, remaining energy, stability period, and system lifespan."

"The sum of alive-nodes in round 3000 is 82 in our proposed protocol whereas in existing methods DETSSEP, ETSSEP, TEDRP, ECDA, EHDT, and NFECG the quantity of alive-node is 71, 35, 40, 38, 25, and 50 respectively."

"In 3000 rounds, the proposed SOSS has 32 J remaining energy whereas DETSSEP, ETSSEP, TEDRP, ECDA, EHDT, and NFECG have 29 J, 22 J, 19 J, 15 J, 11 J, and 10 J respectively in 3000 rounds."

"From the analysis, the proposed method achieved the best performances as compared to other existing approaches in terms of the number of alive nodes, the number of dead nodes, throughput, and the remaining energy."

Conclusion

"A hybrid SOSS for energy-efficient RPs based routing incorporated with multiple MSs in the HWSNs model has been proposed."

"Node centrality, residual energy, and intra-cluster distance are deliberated for the choice of optimal CHs."

"For energy-efficient data collection, multiple MSs are used which ameliorates system lifespan and energy consumption."

"The energy-efficient routing method of the hybrid SOSS algorithm is employed to discover the minimum quantity of RPs and travelling route of MS."

"Further study and research will be needed to evaluate and customize energy-efficient reverse paths for IoT scenarios using mobile nodes and sink."

Functional Materials and Devices by Self-Assembly [61]

This is a machine-generated summary of:

Talapin, Dmitri V.; Engel, Michael; Braun, Paul V.: Functional materials and devices by self-assembly [61].

Published in: MRS Bulletin (2020).

Link to original: https://doi.org/10.1557/mrs.2020.252

If you want to cite the papers, please refer to the original.

For technical reasons we could not place the page where the original quote is coming from.

Abstract-Summary

"The field of self-assembly has moved far beyond early work, where the focus was primarily the resultant beautiful two- and three-dimensional structures, to a focus on forming materials and devices with important properties either otherwise not available, or only available at great cost."

"Materials with unprecedented electronic, photonic, energy-storage, and chemical separation functionalities were created with self-assembly, while at the same time, the ability to form even more complex structures in two and three dimensions has only continued to advance."

"As the field of self-assembly continues to advance, the number of highly functional systems will only continue to grow and make increasingly greater impacts in both the consumer and industrial space."

Introduction

"Small atomic assemblies, aka molecules, remain the most fundamental and important concept in chemistry."

"[62] However, organized structures can form spontaneously, not only from atoms and small molecules, but also from various other types of building blocks."

"Self-assembly creates an opportunity to develop new paradigms for chemistry and material science, where various, typically nanometer-sized, objects with precisely engineered sizes, shapes, compositions, and concomitant properties serve as "meta-atoms" or superatomic building blocks for hierarchically assembled materials and devices."

"Just as atoms combine to form molecules with dramatically different properties than the atomic constituents, self-assembly of "meta-atoms" can create "meta-molecules," and "meta-crystals."

What Is Self-Assembly Good for?

"Self-assembly adds several unique features to our existing toolset of chemical and physical methods for the synthesis and processing of functional materials."

"Self-assembly is particularly useful to synthesize hierarchically organized materials with structures independently engineered on different scales."

"A variety of macromolecules containing two or more covalently bonded blocks of different polymers can be prepared by conventional chemical synthesis."

"The type of self-assembling structure and feature size can be rationally engineered by controlling the block size of individual molecules."

Different Approaches to Classify Self-Assembly Phenomena

"Given the breadth of self-assembly phenomena and materials systems, the classification of these effects can be approached from different angles."

"In dynamic self-assembly, on the other hand, structures or patterns form away from equilibrium."

"Self-assembly is associated with noncovalent interactions, such as van der Waals forces, long-ranged electrostatic, magnetic interactions, and hydrogen bonding."

"[63] However, when the assembling blocks are larger than atoms and small molecules, the interactions can be much more complex than interatomic forces."

"In the first category, the local assembly rules are binary like-dislike type interactions (e.g., between hydrophobic and hydrophilic domains of a polymer backbone) [64]. Even these simple interactions, combined with precise control over size and shape of assembling units, can lead to complex structures."

"Complex ordered structures can form spontaneously even in the absence of any local attractive or repulsive forces between assembling units."

Theoretical and Computational Insights in Self-Assembly

"Self-assembly processes of nanoscale building blocks are founded on statistical mechanics."

"Modeling is best accomplished with computer simulations."

"It is rarely necessary to include quantum mechanical effects explicitly in the modeling process to study self-assembly."

"Quantum effects can become relevant when analyzing physical and chemical properties of the final self-assembled material."

"Ab initio quantum chemistry methods can assist parametrization of coarser simulations with classical force fields, which foremost must reproduce van der Waals force accurately as those are often difficult to estimate and most crucial for self-assembly."

"The most successful applications of self-assembly simulations are structure prediction (local order, mesophases, crystallographic order) and resolving particle dynamics."

"[65, 66] While quantitative theoretical predictions remain difficult with room for future improvement, theory already routinely provides assistance for mechanistic understanding of self-assembly processes, helps improve simulation parameters, and inspires new research directions."

From New Structures to New Functions

"Self-assembly allows combining dissimilar materials into one structure while enhancing the function beyond that of the building blocks."

"The critical role of interfaces becomes the crosscutting theme in self-assembly of functional materials and devices."

"[67] In many cases, self-assembly helped integrate active components; such as semiconductor quantum dots, carbon nanotubes and polymer molecules in the complete device structure."

"All of these devices rely on efficient transport of charge carriers, electrons or ions, through self-assembled materials."

"Fcc superlattices self-assembled from spherical silica or poly(methyl methacrylate) (PMMA) particles with a diameter of hundreds of nanometers to micrometers exhibit the properties of photonic crystals."

"[68] However, the refractive indexes of SiO_2 and PMMA are insufficient to develop a complete photonic bandgap, while high-index materials, such as TiO_2 or Si, could not be prepared as monodisperse spheres suitable for self-assembly into long-range ordered superlattices."

From Function to Market

"The preceding sections, self-assembled structures can show not only unprecedented structural motifs on previously inaccessible length scales in both two and three dimensions, but importantly, also provide materials with unique physical and chemical properties."

"Long-term stability and environmental concerns must be addressed for the successful adaption of self-assembled materials by the marketplace."

"At this relatively early stage, several self-assembled materials and devices have been integrated in consumer products or implemented in large-scale manufacturing processes and more are currently on a commercialization pathway."

"One of the obstacles that complicates adoption of self-assembled materials by the nanoelectronics community is structural defects arising from small local variations in process parameters."

"Similar analysis can be applied to many other application areas for self-assembled materials."

"[69] It is only a matter of time until there is an increase in the numbers of materials and devices with self-assembled components in the market."

Future Directions for Self-Assembly

"We also expect the development of advanced computational models and tools with good predictive power for the rational design of functional materials by self-assembly."

"Equilibrium assembly represents just a subset of possible self-organization phenomena."

"All living systems, for example, rely on complex networks of nonequilibrium self-assembly."

"[70] This introduces new properties, such as odd elasticity [71] and odd viscosity [72] that are forbidden in static materials, and calls for different theoretical frameworks for describing and classifying nonequilibrium self-assembly phenomena."

"We can only speculate about what applications and technologies will emerge once we develop a better understanding of physical and chemical principles of nonequilibrium self-assembly."

"The second area of huge potential relates to the coupling strength of the components in self-assembled materials."

"[73, 74] The quality of self-assembled materials only recently approached levels needed to observe such effects."

"The time may be just right to launch systematic investigations and engineering of quantum phenomena in self-assembled materials."

In this Issue

"This issue of MRS Bulletin combines articles written by leaders in key areas of fundamental and translational research on self-assembly for functional materials and devices."

"Three articles in this issue cover application-driven implementations of self-assembly."

"The article by H. Chen and others [75] covers recent developments of self-assembly for making better batteries and supercapacitors."

"The Kagan and others, an article discusses the applications of self-assembled materials and self-assembly methods for a plethora of electronic devices, from mainstream CMOS nanofabrication to new devices with unusual form factors, such as flexible and stretchable sensors."

"[76] This forward-looking contribution emphasizes that self-assembly has room for further evolution and expansion into the world of quantum materials and devices."

Conclusion

"Two decades of active research on self-assembly has delivered materials with unprecedented nanoscale structures in both two and three dimensions."

"In early work, the focus was primarily the nanostructure of the self-assembled materials."

"The current state of the field, as covered in this MRS Bulletin issue, strongly suggests that self-assembly is making significant strides toward application in nanoelectronics [77], photonics [76], energy storage [75], chemical separations [78], and as a path to form complex structures."

Water Cycle Algorithm Perspective on Energy Constraints in WSN [79]

This is a machine-generated summary of:

Tiwari, Sudhanshu; Kumar, Gaurav; Raj, Ayush; Prateek, None; Arya, Rajeev: Water cycle algorithm perspective on energy constraints in WSN [79].

Published in: International Journal of System Assurance Engineering and Management (2019).

Link to original: https://doi.org/10.1007/s13198-019-00784-y

If you want to cite the papers, please refer to the original.

For technical reasons we could not place the page where the original quote is coming from.

Abstract-Summary

"WCA has been applied independently to three energy models of WSN, with a view to optimize the energy spent on data communication in the nodes."

"By taking different sets of node population sizes, the energy performance is analyzed from the point of view of the percentage reduction of energy depleted, and the number of iterations needed to achieve the optimum value."

"The implementation renders an energy reduction of up to 46.24% in the sensor nodes, which is a promising outcome, and must be investigated further."

Introduction

"Some of the constraints in WSNs, such as the processing power of the nodes, storage capacity, data management, routing in difficult terrains, optimization of energy consumption, extending communication range need to be improved (Pantazis and others [80]; Younis and Akkaya [81])."

"Of all these, energy consumption optimization is a topic of concern and a key area of research in WSN (Anastasi and others [82]; Karahan and others [83]; Wsns and others [84]; Mini and Loureiro [85]; Karim and Zeadally [86])."

"It has been found through researches that more energy is consumed in data communication as compared to sensing and data processing in WSN (Yick and others [87]; Snajder and others [88]; Raza and others [89]; Norouzi and Zaim [90]; Peiravi and others [91])."

"It is important to devise optimization techniques for minimization of energy consumption in communication process in WSNs."

"Various research reports show that nature inspired artificial intelligence (AI) methods are being utilized for the energy optimization in WSN."

"This is the main motivation behind the present study wherein WCA optimization technique has been applied to three energy models that compute the communication energy consumption perspective of WSNs."

WSN Energy Models

"Sensor nodes must be designed in such a way that they must utilize minimum energy and thus maximize the battery life."

"In a simulation study (Zhang and others [92]) the values were taken as $E_{elec} = 50$ nJ/bit, $\epsilon_{fs} = 10$ pJ/bit/m^2, $\epsilon_{mp} = 0.0013$ pJ/bit/m^4, $d_0 = 87$ m. The present study utilizes the three energy models mentioned above because of their proximity to estimate the wireless sensor node energy consumption."

"WCA method has been applied to the above energy models to obtain the optimum conditions for the minimum consumption of energy in WSN."

Water Cycle Optimization Algorithm (WCA)

"During this transition, the quantity of aggregation may change, thus forming water bodies of varying importance and thickness (Water-drops, streams, rivers, sea)."

"Since sea is a huge resource in terms of water volume, the algorithm assigns the greatest importance to it, followed by the river and lastly the streams."

"At each step, the importance calculation is done, which iteratively determines the best possible value of the water-body, and finally renames it to 'sea'."

"The rest of the nodes are sorted in increasing order of their energy depletion cost as efficient nodes lie near the cluster head (rivers flowing into the sea), while lower efficiency ones lie farther away from the head node (Streams farthest from the sea)."

"With each iteration, the energy depletion rate is re-calculated for each node, and accordingly some nodes communicate directly with the head node (streams flowing directly into the sea)."

Simulation Results and Discussion

"The minimum energy obtained for energy model 1 is -4.0028×10^{-8} J with value of M as 0.8 and packet size of 1000 bits over 9 iterations."

"In the energy model 3 the minimum energy obtained is 3.4895×10^{-4} J with the value of distance (d) as 87 m and packet size of 2000 bits."

"On comparing the three energy models E1, E2 and E3 for each value of population it is observed that, E2 energy model works better for smaller population size of nodes."

"This efficiency reduces with the increase in the population size, i.e. for larger population, the percentage of energy reduced is in single digit percentage value except for the first energy model in which the percentage value is significantly low."

"If an energy model takes a greater number of iterations to reach optimum value for a particular population size, keeping other parameters constant, there is stronger impact on energy reduction."

Conclusion

"Water cycle algorithm (WCA) has been applied to minimize the value during data communication."

"Three energy models are chosen in the present study and optimization technique is applied on them."

"WCA implementation is limited to the energy models of WSNs."

"Future studies shall target applying advanced techniques like neural networks, machine learning, deep learning etc. on the models to try to find better results and to replace the algorithms that are applied for optimization of various problems solved in electronics with WCA and compare the results."

Novel Enhanced LoRaWAN Framework for Smart Home Remote Control Security [93]

This is a machine-generated summary of:

Naoui, Sarra; Elhdhili, Mohamed Elhoucine; Azouz Saidane, Leila: Novel Enhanced LoRaWAN Framework for Smart Home Remote Control Security [93].

Published in: Wireless Personal Communications (2019).

Link to original: https://doi.org/10.1007/s11277-019-06832-x

If you want to cite the papers, please refer to the original.

For technical reasons we could not place the page where the original quote is coming from.

Abstract-Summary

"Among the most promising smart home applications, we distinguish the remote services control, that requires a secure and long range communication."

"LoRaWAN is considered as a rising technology providing a long range and secure communication."

"Some LoRaWAN security vulnerabilities have been identified in the literature."

"In this paper, we propose a novel enhanced LoRaWAN solution that we adapt to secure the smart home remote control."

Introduction

"Although LoRaWAN has lightweight security measures compared to other LPWA technologies, it also has some security vulnerabilities that may be exploited by attackers to compromise the network security."

"In 2017, Naoui and others [94] identified more other possible attacks on LoRaWAN such as the network server attack, DoS attack, and compromised-key attack."

"The same authors [95] studied the LoRaWAN security vulnerabilities and disclosed that LoRa transmissions are prone also to physical and replay attacks."

"Lee and others [96] studied the risk of a bit-flipping attack in LoRaWAN network and proposed a shuffling method as a countermeasure to avoid it."

"In this paper, we propose a novel enhanced LoRaWAN solution that solves the security vulnerabilities present in basic LoRaWAN and related works."

"In a second part, we propose to apply our novel LoRaWAN solution to the smart home remote control application."

LoRaWAN Specification

"The "join accept" is secured using the AppKey as follows: Once the device joins the network successfully, it computes the two session keys, a NwkSKey used to secure the communication between the end device and the network server, and the AppSKey used to secure the communication between the end device and the application server: Activation By Personalization (ABP): the end device is activated automatically without going through a join procedure as it is equipped with its address DevAddr and the two session keys NwkSKey and AppSKey."

"It is significant to note that after an end device resetting or a join procedure, the frame counters on the end device and the network server sides are reset to 0 [97]."

"OTAA method introduces the exchange of two messages (Join Request, Join Accept) between the end device and the network server."

Novel Enhanced LoRaWAN

"Upon receiving the message, the network server computes MIC2 and checks the message integrity, once valid, it sends the ciphertext with MIC1 to the application server, that decrypts the message and checks its integrity using MIC1. To avoid physical attacks, we propose to employ a secure hardware element in the end device in order to preserve the security credentials."

"Outside the smart home, we have a LoRaWAN Cloud network server that verifies the integrity of the messages exchanged between the end devices and the user application, and an application server that presents the user smart home application device (e.g. smartphone, tablet, home control interface ...)."

"Application Server For data visualization and for interaction between the user and its smart home devices, end user devices are used (exemple: HCI, user smartphone, user tablet)."

Formal Verification, Security and Performance Analysis

"Proof (1) Resist masquerade attack: An attacker cannot masquerade as a legal entity among the HCI and an end device A in order to join the smart home network."

"He cannot send a fake "Join Request 1″ to the HCI because the attacker must get the AppKey shared between the end device A and the HCI, as well as the AppEUI which is only known to the end device A and the application server. (2) Resist message-forgery attack: We assume that an attacker may capture the previous messages exchanged during the join procedure between the HCI and an end device A. He cannot breach the messages integrity because they are encrypted using AppKey and signed using MIC values."

"Proposition: The proposed solution is secure against replay attack and known-key attack."

"Our solution is based on the end device to application server communication and conversely, where each end device stores its unique keys shared with the network server and the application server."

Conclusion

"We proposed a novel enhanced LoRaWAN solution that we apply to the smart home context to get a new smart home remote control security framework."

"The formal analysis using Scyther tool proved that the proposed scheme performs authentication and confidentiality, and also security goals as expected."

"The performance analysis demonstrated that session keys are established in a lightweight and efficient way compared to existing solutions."

References

1. Korkmaz M (2022) A novel method for SoC estimation of lithium-ion batteries based on previous covariance matrices and variable ECM parameters. Electr Eng. https://doi.org/10.1007/s00202-022-01692-4
2. Rivera-Barrera JP, Muñoz-Galeano N, Sarmiento-Maldonado HO (2017) SoC estimation for lithium-ion batteries: review and future challenges. Electronics 6(4):102
3. Meena T, Sangam RS (2022) Generic potential field based distributed node coordination in flying adhoc network (FANET). J Ambient Intell Humaniz Comput. https://doi.org/10.1007/s12652-022-03767-3
4. Khan A, Aftab F, Zhang Z (2019) Self-organization based clustering scheme for FANETs using glowworm swarm optimization. Phys Commun 36:100769
5. Razzaq S, Xydeas C, Everett ME, Mahmood A, Alquthami T (2018) Three-dimensional UAV routing with deconfliction. IEEE Access 6:21536–21551
6. Radmanesh M, Kumar M, Guentert PH, Sarim M (2018) Overview of path-planning and obstacle avoidance algorithms for UAVs: a comparative study. Unmanned Syst 06(02):95–118
7. Anicho O, Charlesworth PB, Baicher G, Nagar A (2018) Integrating routing schemes and platform autonomy algorithms for UAV Ad-hoc & infrastructure based networks. In: 28th international telecommunication networks and applications conference (ITNAC)
8. Li D, Du Y (2007) Artificial intelligence with uncertainty. Chapman and Hall/CRC, London, pp 193–211
9. Sha W, Guo Y, Cheng D, Han Q, Lou P, Guan M, Tang S, Zhang X, Lu S, Cheng S, Cao Y-C (2022) Degradation mechanism analysis of $LiNi_{0.5}Co_{0.2}Mn_{0.3}O_2$ single crystal cathode materials through machine learning. npj Comput Mater. https://doi.org/10.1038/s41524-022-00905-5
10. Khanum F, Louback E, Duperly F, Jenkins C, Kollmeyer PJ, Emadi A (2021) A Kalman filter based battery state of charge estimation MATLAB function. In: 2021 IEEE transportation electrification conference & expo (ITEC). IEEE, pp 484–489
11. Chung D-W, Ko J-H, Yoon K-Y (2022) State-of-charge estimation of Lithium-ion batteries using LSTM deep learning method. Journal of electrical. Eng Technol. https://doi.org/10.1007/s42835-021-00954-8
12. Lu L, Han X, Li J, Hua J, Ouyang M (2013) A review on the key issues for lithium-ion battery management in electric vehicles. J Power Sources 226:272–288

13. Hannan MA, Lipu MH, Hussain A, Mohamed A (2017) A review of lithium-ion battery state of charge estimation and management system in electric vehicle applications: challenges and recommendations. Renew Sust Energ Rev 78:834–854
14. Yang F, Li W, Li C, Miao Q (2019) State-of-charge estimation of lithium-ion batteries based on gated recurrent neural network. Energy 2:196
15. Liu P, Qiu X, Huang X (2016) Recurrent neural network for text classification with multi-task learning. arXiv. preprint arXiv:1605.05101
16. Zhang X, Zhou J, Lu J, Shen L (2022) Interpretable learning of voltage for electrode design of multivalent metal-ion batteries. npj Comput Mater. https://doi.org/10.1038/s41524-022-00858-9
17. Joshi RP et al (2019) Machine learning the voltage of electrode materials in metal-ion batteries. ACS Appl Mater Inter 11:18494–18503
18. Moses IA et al (2021) Machine learning screening of metal-ion battery electrode materials. ACS Appl Mater Interfaces 13:53355–53362
19. Zhou F, Cococcioni M, Marianetti CA, Morgan D, Ceder G (2004) First-principles prediction of redox potentials in transition-metal compounds with LDA+U. Phys Rev B Condens Matter Mater Phys 70:235121
20. Jain A et al (2013) Commentary: the materials project: a materials genome approach to accelerating materials innovation. APL Mater 1:011002
21. Chu S, Cui Y, Liu N (2017) The path towards sustainable energy. Nat Mater 16:16–22
22. Rendón MA, Sánchez R, Carlos D, Josselyn GM, Anzai AH (2021) Aircraft hybrid-electric propulsion: development trends, challenges and opportunities. J Control Autom Electr Syst. https://doi.org/10.1007/s40313-021-00740-x
23. Zhang M, Eastham F, Yuan W (2016) Design and modeling of 2G HTS armature winding for electric aircraft propulsion applications. IEEE Trans Appl Supercond 26(3):1–5. https://doi.org/10.1109/TASC.2016.2539548
24. Sliwinski J, Gardi A, Marino M, Sabatini R (2017) Hybrid-electric propulsion integration in unmanned aircraft. Energy 140:1407–1416. https://doi.org/10.1016/j.energy.2017.05.183
25. IATA (2019). Aircraft Technology Roadmap to 2050. https://www.iata.org/contentassets/8d19e716636a47c184e7221c77563c93/technology20roadmap20to20205020no20foreword.pdf
26. Liu C (2013) Turboelectric distributed propulsion system modelling. PhD thesis. Cranfield University
27. National Academies of Sciences, Engineering and Medicine (2016) Commercial aircraft propulsion and energy systems research: reducing global carbon emissions. The National Academies of sciences, DC. https://doi.org/10.17226/23490
28. Brelje BJ, Martins JR (2019) Electric, hybrid, and turboelectric fixed-wing aircraft: a review of concepts, models, and design approaches. Prog Aerosp Sci 104:1–19. https://doi.org/10.1016/j.paerosci.2018.06.004
29. HECARRUS (2020) WP1:"Efficiency and TRL of each component of the powertrai" D1.1: "State-of-the-art technologies on research and developments underway in the field of alternative propulsion architecture". In: Horizon 2020 / Clean Sky JU https://ec.europa.eu/research/participants/documents/
30. Wright Electric (2020) Lower cost, quieter flight, cleaner future. https://weflywright.com/
31. Misra A (2018) Energy storage for electrified aircraft: the need for better batteries, fuel cells, and supercapacitors. IEEE Electrif Mag 6(3):54–61
32. Continental, A T. (2019). Continental aerospace Technologies™ announces strategic partnership with VerdeGo Aero™ to develop hybrid-electric aerospace powertrains. https://continentaldiesel.com/news-events/continental-aerospace-technologiesannounces-strategic-partnership-with-verdego-aero-to-develop-hybrid-electric-aerospace-powertrains/
33. EEPower. (2020). VerdeGo aero aiming for hybrid-electric autonomous personal air taxi. https://eepower.com/news/verdego-aero-aiming-for-hybrid-electric-autonomous-personal-air-taxi/#

34. Borer NK, Patterson MD, Viken JK, Moore MD, Bevirt J, Stoll AM, Gibson AR (2016) Design and performance of the NASA SCEPTOR distributed electric propulsion flight demonstrator. In: 16th AIAA aviation technology, integration, and operations conference, p 3920
35. Patterson MD, Derlaga JM, Borer NK (2016) High-lift propeller system configuration selection for NASA'S SCEPTOR distributed electric propulsion flight demonstrator. In: 16th AIAA aviation technology, integration, and operations conference. https://doi.org/10.2514/6.2016-3922
36. González Espasandín Ó, Leo TJ, Navarro-Arévalo E (2014) Fuel cells: a real option for unmanned aerial vehicles propulsion. Sci World J 2014:1–12. https://doi.org/10.1155/2014/497642
37. Sziroczak D, Jankovics I, Gal I, Rohacs D (2020) Conceptual design of small aircraft with hybrid-electric propulsion systems. Energy:117937
38. Kim HD, Brown GV, Felder JL (2008) Distributed turboelectric propulsion for hybrid wing body aircraft. In: International powered lift conference Royal Aeronautical Society. NASA Glenn Research Center, London, pp 22–24
39. Yang S, Page M, Smetak EJ (2018b) Achievement of NASA new aviation horizons n+2goals with a blended-wing-body x-plane designed for the regional jet and single-aisle jet markets. In: AIAA aerospace sciences meeting. American Institute of Aeronautics and Astronautics. https://doi.org/10.2514/6.2018-0521
40. Boeing. (2017) Blended Wing Body Goes with the Flow in New Visualization Test. http://www.boeing.com/features/2017/05/blended-wing-body-05-17.page
41. Harrison NA, Gatlin GM, Viken SA, Beyar M, Dickey ED, Hoffman K, Reichenbach EY (2020) Development of an efficient m=0.80 transonic truss-braced wing aircraft. In: AIAA Scitech 2020 Forum. American Institute of Aeronautics and Astronautics. https://doi.org/10.2514/6.2020-0011
42. Bradley M, Droney C (2012) Subsonic ultra green aircraft research phase ii: n+ 4 advanced concept development. techreport CR-2012-217556, NASA
43. Van Oppen J, Vos R, Boneschansker I, de Koo M (2020) Flying-V. Flying long distances energy-efficiently. https://www.tudelft.nl/en/ae/flying-v/
44. Boeing (2019) Spreading our wings: Boeing unveils new Transonic Truss-Braced Wing. https://www.boeing.com/features/2019/01/spreading-our-wings-01-19.page
45. Tasca AL, Cipolla V, Abu SK, Puccini M (2021) Innovative box-wing aircraft: emissions and climate change. Sustain For 13(6):3282
46. Palaia (2021) THEA-CODE: a design tool for the conceptual design of hybrid-electric aircraft with conventional or unconventional airframe configurations.22, pp 19. EDP Sciences
47. Corduan M, Boll M, Bause R, Oomen MP, Filipenko M, Noe M (2020) Topology comparison of superconducting ac machines for hybrid electric aircraft. IEEE Trans Appl Supercond 30(2):1–10
48. Gao XZ, Hou ZX, Guo Z, Chen XQ (2015) Reviews of methods to extract and store energy for solar-powered aircraft. Renew Sust Energ Rev 44:96–108
49. Vetters DK, Karam M, Fulayter RD (2014) Ultra high bypass ratio turbofan engine. US Patent App. 14/101,438
50. GE Aviation (2020) Introducing GE's Catalyst™ advanced turboprop engine. https://www.geaviation.com/bga/engines/ge-catalyst
51. Chen W, Pan J, Fan B, Liu Y, Peter O (2017) Effect of injection strategy on fuel-air mixing and combustion process in a direct injection diesel rotary engine (DI-DRE). Energy Convers Manag 154:68–80
52. Yang J, Ji C, Wang S, Wang D, Shi C, Ma Z et al (2018a) Numerical study of hydrogen direct injection strategy on mixture formation and combustion process in a partially premixed gasoline Wankel rotary engine. Energy Convers Manag 176:184–193
53. Bongermino E, Mastrorocco F, Tomaselli M, Monopoli VG, Naso D (2017) Model and energy management system for a parallel hybrid electric unmanned aerial vehicle. In: 2017 IEEE 26th international symposium on industrial electronics (ISIE), pp 1868–1873. https://doi.org/10.1109/ISIE.2017.8001534

54. Gupta P, Tripathi S, Samayveer S (2021) Energy efficient rendezvous points based routing technique using multiple mobile sink in heterogeneous wireless sensor networks. Wirel Netw. https://doi.org/10.1007/s11276-021-02714-y
55. Wang J, Cao J, Ji S, Park JH (2017) Energy-efficient cluster-based dynamic routes adjustment approach for wireless sensor networks with mobile sinks. J Supercomput 73(7):3277–3290
56. Sharma S, Puthal D, Jena SK, Zomaya AY, Ranjan R (2017) Rendezvous based routing protocol for wireless sensor networks with mobile sink. J Supercomput 73(3):1168–1188
57. Xie J, Zhang B, Zhang C (2020) A novel relay node placement and energy efficient routing method for heterogeneous wireless sensor networks. IEEE Access
58. Verma A, Kumar S, Gautam PR, Kumar A (2020) Stable energy-efficient routing algorithm for dynamic heterogeneous wireless sensor networks. In: Advances in VLSI, communication, and signal processing. Springer, Singapore, pp 151–160
59. Rani R, Kakkar D, Kakkar P, Raman A (2019) Distance based enhanced threshold sensitive stable election routing protocol for heterogeneous wireless sensor network. In: Computational intelligence in sensor networks. Springer, Berlin, pp 101–122
60. Manchanda R, Sharma K (2021) A novel framework for energy-efficient compressive data gathering in heterogeneous wireless sensor network. Int J Commun Syst 34(3):e4677
61. Talapin DV, Engel M, Braun PV (2020) Functional materials and devices by self-assembly. MRS Bull. https://doi.org/10.1557/mrs.2020.252
62. Pauling L (1960) The nature of the chemical bond and the structure of molecules and crystals: an introduction to modern structural chemistry. Cornell University Press, Ithaca, NY
63. Boles MA, Engel M, Talapin DV (2016) Self-Assembly of Colloidal Nanocrystals: From Intricate Structures to Functional Materials. Chem Rev 116:11220. https://pubs.acs.org/doi/10.1021/acs.chemrev.6b00196
64. Förster S, Plantenberg T (2002) From Self-Organizing Polymers to Nanohybrid and Biomaterials. Angew Chem Int Ed 41:688. https://onlinelibrary.wiley.com/doi/10.1002/1521-3773(20020301)41:5%3C688::AID-ANIE688%3E3.0.CO;2-3
65. Akcora P, Liu H, Kumar SK, Moll J, Li Y, Benicewicz BC, Schadler LS, Acehan D, Panagiotopoulos AZ, Pryamitsyn V, Ganesan V, Ilavsky J, Thiyagarajan P, Colby RH, Douglas JF (2009) Anisotropic self-assembly of spherical polymer-grafted nanoparticles. Nat Mater 8:354. https://www.nature.com/articles/nmat2404
66. Travesset A (2017) Topological structure prediction in binary nanoparticle superlattices. Soft Matter 13:147. https://doi.org/10.1039/C6SM00713A
67. Kagan CR, Lifshitz E, Sargent EH, Talapin DV (2016) Building devices from colloidal quantum dots. Science 353:aac5523. https://www.science.org/doi/10.1126/science.aac5523
68. Joannopoulos JD, Johnson SG, Winn JN, Meade RD (2008) Photonic crystals: molding the flow of light, 2nd edn. Princeton University Press, Princeton, NJ
69. aculon.com, "Hydrophobic, Oleophobic, and Hydrophilic Surface Treatments," (accessed August 25, 2020)
70. Shaebani MR, Wysocki A, Winkler RG, Gompper G, Rieger H (2020) Computational models for active matter. Nat Rev Phys 2:181. https://www.nature.com/articles/s42254-020-0152-1
71. Scheibner C, Souslov A, Banerjee D, Surówka P, Irvine WTM, Vitelli V (2020) Odd elasticity. Nat Phys 16:475. https://www.nature.com/articles/s41567-020-0795-y
72. Banerjee D, Souslov A, Abanov AG, Vitelli V (2017) Odd viscosity in chiral active fluids. Nat Commun 8:1573. https://www.nature.com/articles/s41467-017-01378-7
73. Kagan CR, Murray CB (2015) Charge transport in strongly coupled quantum dot solids. Nat Nanotechnol 10:1013. https://www.nature.com/articles/nnano.2015.247
74. Mueller NS, Okamura Y, Vieira BGM, Juergensen S, Lange H, Barros EB, Schulz F, Reich S (2020) Deep strong light–matter coupling in plasmonic nanoparticle crystals. Nature 583:780. https://www.nature.com/articles/s41586-020-2508-1
75. Chen H, Gao C, Xu H (2021) Advanced Graphene Materials for Sodium/ Potassium/ Aluminum-Ion Batteries. ACS Materials Lett. 3(8):1221–237. https://doi.org/10.1021/acsmaterialslett.1c00280

76. Rainò G, Utzat H, Bawendi MG, Kovalenko MV (2020) Superradiant emission from self-assembled light emitters: From molecules to quantum dots. MRS Bull 45(10):841. https://doi.org/10.1557/mrs.2020.250
77. Kagan CR, Hyeon T, Kim D-H, Ruiz R, Tung MC, Wong HSP (2020) Self-assembly for electronics. MRS Bull 45(10):807. https://link.springer.com/article/10.1557/mrs.2020.248
78. Chen F, Day GS, Zhou H-C (2020) Separation using self-assembled materials. MRS Bull 45(10):823. https://doi.org/10.1557/mrs.2020.246
79. Tiwari S, Kumar G, Raj A, Prateek N, Arya R (2019) Water cycle algorithm perspective on energy constraints in WSN. Int J Syst Assur Eng Manag. https://doi.org/10.1007/s13198-019-00784-y
80. Pantazis NA, Nikolidakis SA, Vergados DD, Member S (2013) Energy-efficient routing protocols in wireless sensor networks: a survey. IEEE Commun Surv Tutor 15:551–591. https://doi.org/10.1109/SURV.2012.062612.00084
81. Younis M, Akkaya K (2008) Strategies and techniques for node placement in wireless sensor networks: a survey. Ad Hoc Netw 6:621–655. https://doi.org/10.1016/j.adhoc.2007.05.003
82. Anastasi G, Conti M, Di M, Passarella A (2009) Energy conservation in wireless sensor networks: a survey. Ad Hoc Netw 7:537–568. https://doi.org/10.1016/j.adhoc.2008.06.003
83. Karahan A, Erturk I, Atmaca S, Cakici S (2014) Effects of transmit-based and receive-based slot allocation strategies on energy efficiency in WSN MACs. Ad Hoc Netw 13:404–413. https://doi.org/10.1016/j.adhoc.2013.09.001
84. Wsns SL, Chidean MI, Morgado E et al (2016) Energy efficiency and quality of data reconstruction through data-coupled clustering. IEEE Sens Netw 16:5010–5020. https://doi.org/10.1109/JSEN.2016.2551466
85. Mini RAF, Loureiro AAF (2009) Energy in wireless sensor networks. In: Garbinato B, Miranda H, Rodrigues L (eds) Middleware for network eccentric and mobile applications. Springer, Berlin, pp 3–24. https://doi.org/10.1007/978-3-540-89707-1_1
86. Karim F, Zeadally S (2016) Energy harvesting in wireless sensor networks: a comprehensive review. Renew Sust Energ Rev 55:1041–1054. https://doi.org/10.1016/j.rser.2015.11.010
87. Yick J, Mukherjee B, Ghosal D (2008) Wireless sensor network survey. Comput Netw 52:2292–2330. https://doi.org/10.1016/j.comnet.2008.04.002
88. Snajder B, Jelicic V, Kalafatic Z, Bilas V (2016) Wireless sensor node modelling for energy efficiency analysis in data-intensive periodic monitoring. Ad Hoc Netw 49:29–41. https://doi.org/10.1016/j.adhoc.2016.06.004
89. Raza U, Bogliolo A, Freschi V et al (2016) A two-prong approach to energy-efficient WSNs: wake-up receivers plus dedicated, model-based sensing. Ad Hoc Netw 45:1–12. https://doi.org/10.1016/j.adhoc.2016.03.005
90. Norouzi A, Zaim AH (2014) Genetic algorithm application in optimization of wireless sensor networks. Sci World J. https://doi.org/10.1155/2014/286575
91. Peiravi A, Mashhadi HR, Javadi SH (2013) An optimal energy-efficient clustering method in wireless sensor networks using multi-objective genetic algorithm. Int J Commun Syst 26:114–126. https://doi.org/10.1002/dac.1336
92. Zhang H, Zhang S, Bu W (2014) A clustering routing protocol for energy balance of wireless sensor network based on simulated annealing and genetic algorithm. Int J Hybrid Inf Technol 7:71–82
93. Naoui S, Elhdhili ME, Saidane LA (2019) Novel enhanced LoRaWAN framework for smart home remote control security. Wirel Pers Commun. https://doi.org/10.1007/s11277-019-06832-x
94. Yang X (2017) LoRaWAN: vulnerability analysis and practical exploitation. Delft University of Technology. URL: https://repository.tudelft.nl/islandora/object/uuid:87730790-6166-4424-9d82-8fe815733f1e?collection=education
95. Aras E et al (2017) Exploring the security vulnerabilities of LoRa. In: 2017 3rd IEEE international conference on cybernetics (CYBCONF). IEEE
96. Lee J et al (2017) Risk analysis and countermeasure for bit-flipping attack in LoRaWAN. In: 2017 international conference on information networking (ICOIN). IEEE
97. Alliance L (2016) LoRaWAN specification (V1. 0.2). [2017-06-10]. https://www.lora-alliance.org/Contact

Conclusion by the Editor

After reading all the chapters from this book, it has taught readers to learn the depth of knowledge and practical applications of Quantum Artificial Intelligence (QAI). This book ensures that the learners get a feeling of how to design, develop, prototype, and commercialize the products or solutions towards QAI.

Chapter 1 deals with the quantum information processing, the arithmetic operations of quantum information science, Qubit analysis and transmission of information.

Chapter 2 deals with quantum programming which enables the learners to compute quantum array gates and quantum logical programming.

Chapter 3 deals with the solved and unsolved problems in post-quantum cryptography. Quantum computing is a threat to conventional crypto algorithms and protocols, this chapter has illustrated to learners.

Chapter 4 deals with the simulation of quantum and artificial systems which envisions the learners to scale up 'new dimensions and its mixture' in both subjects: Quantum systems and Artificial Intelligence systems.

Chapter 5 deals with quantum machine learning which deals with the data science models of the voluminous and variety of dataset and event logs.

At the end of all the Chaps. 1–5 the readers will have a wide scope on the industrial importance and usage of QAI. Chapter 6 deals with the applications of QAI covering multiple domains.

In the next version of the book, the author would like to bring out an emerging field 'QAI systems and methods' through Quantum Open AI software tools. Happy to receive the feedback of this book through the publisher. Hoping for the best!

V. Natarajan (ed.), *Quantum Artificial Intelligence*,
https://doi.org/10.1007/978-981-96-5051-4

The manufacturer's authorised representative in the EU is Springer Nature Customer Service Centre GmbH, Europaplatz 3, 69115 Heidelberg, Germany. If you have any concerns regarding our products, please contact ProductSafety@springernature.com

Printed and bound by CPI Group (UK) Ltd, Croydon, CR0 4YY
01/10/2026
02225923-0011